The Science of NeuroMarketing

Decoding Buyers Behavior:

25 Powerful Strategies to
Maximize Revenue by Mastering the
Customers Buying Conversation

Andy Broadaway

Abundant Press www.AbundantPress.com

Ordering Information & Quantity Sales: Special discounts are available on quantity purchases by corporations, associations, and others. For details, contact the publisher. Orders by U.S. & International trade bookstores and wholesalers at: info@abundantpress.com. Printed in the United States of America. Library of Congress-in-Publication Data. Please note: If you find errors, like spelling or grammar, please send us an email.

Title: The Science of NeuroMarketing

Sub-title: 25 Powerful Strategies to Maximize Revenue by Mastering The Customer Buying Conversation

Author: Andy Broadaway

1. The main category of the book — Business, Leadership. Marketing, Sales

ISBN: 978-1-948287-54-8

To Get FREE Access Additional Bonuses Visit:

NeuromarketingBook.com

Disclaimer

Although the author(s) and publisher have made every effort to ensure that the information in this book was correct at press time. While we try to keep the information up-to-date and accurate, there are no representations or warranties, express or implied, about the completeness, accuracy, reliability, suitability or availability with respect to the information. The author and publisher are NOT providing tax, legal, financial or investment advice, do not assume and hereby disclaim any liability to any party for any loss, damage, or disruption caused by errors or omissions, whether such errors or omissions result from negligence, accident, or any other cause.

Nothing else contained in this book should be used or construed as an offer to sell, a solicitation of an offer to buy, or a recommendation for any security. Nor is it intended as investment, tax, financial or legal advice. Investors should seek such professional advice for their particular situation.

This book is for informational purposes only. This book should not be considered a solicitation, offer or recommendation for the purchase or sale of any securities or other financial products and services discussed herein. Readers of this book will not be considered clients of Andy Broadaway just by virtue of access to this book. Information contained herein is not intended for persons in any jurisdiction where such distribution or use would be contrary to the laws or regulations of that jurisdiction. Readers should not construe any discussion or information contained herein as personalized advice from Andy Broadaway. Neither our information providers nor we shall be liable for any errors or inaccuracies, regardless of cause, or the lack of timeliness of, or for any delay or loss of income.

Certain information contained in this book is derived from sources that Andy Broadaway and/or believes to be reliable; however, Andy Broadaway and/or does not guarantee the accuracy or timeliness of such information and assumes no liability for any resulting damages. Readers should seek advice regarding the appropriateness of investing in any securities or other financial instruments referred to in this book, website or any other message received from Andy Broadaway and should understand that statements regarding future prospects of these or other financial products may not be realized.

Legal Disclosure: You are hereby advised that Andy Broadaway is not a financial advisor and is NOT providing legal or tax advice. Nothing in this book or its attachments should be interpreted by you as legal advice. For legal advice and all legal related matters, Andy Broadaway recommends that you seek the advice of a qualified attorney licensed in your state or jurisdiction.

Foreword by Joseph McClendon III

When I first came across the concept of Neuro-Linguistic Programming (NLP), it felt like a breakthrough—a way to unlock human potential and tap into the psychological drivers behind decision-making. Over the years, I've seen firsthand how the right techniques, rooted in an understanding of human behavior, can make the difference between average results and true success. Whether it's inspiring someone to take action or guiding individuals to transform their lives, the tools NLP provides are extraordinary.

I've spent my career helping people and organizations shift their mindset to achieve greater performance, and one thing has always stood out to me: understanding how people think is crucial if you want to influence them. That's why *The Science of NeuroMarketing* is such a valuable resource. Andy Broadaway has masterfully combined powerful NLP principles with marketing strategies that unlock the mysteries of consumer behavior.

Throughout this book, you'll discover how to tap into the deepest layers of human behavior to craft messages that resonate on a subconscious level. Andy walks you through actionable techniques, like anchoring, scarcity, and social proof, that can be applied to create urgency, build trust, and elevate your brand. These aren't just theoretical concepts—they're real-world strategies you can implement today to see immediate results.

The beauty of Andy's approach is how he breaks down complex ideas into practical steps that anyone can follow, whether you're new to marketing or a seasoned professional. His insights on emotionally driven content are particularly compelling. In today's world, where consumers are bombarded with information from all sides, it's not enough to simply sell a product.

You must craft authentic, emotionally resonant messages that build deeper connections with your audience. That's exactly what this book teaches.

In my own work, I've always emphasized the importance of authenticity and personalization in communication. People are naturally drawn to messages that feel genuine and relatable. Andy not only highlights the power of authenticity in marketing but also provides you with the tools to infuse it into your digital presence, whether through social media, email, or content marketing.

As you embark on this journey, remember that marketing is as much about building relationships as it is about conversions. The techniques outlined in this book—grounded in the science of NLP—offer a roadmap to mastering both. These strategies will help you unlock the psychology behind consumer choices, motivate action, and ultimately, transform curious visitors into loyal customers.

This book is more than just a guide—it's a powerful tool to elevate your marketing game, increase conversions, and create lasting customer connections. Andy Broadaway has given us a gift in this work, and I encourage you to dive deep into every chapter, absorb the wisdom, and apply the strategies.

But, what really sets this book apart is Andy's ability to blend real-world marketing strategies with the latest advances in NLP. The result is a comprehensive guide that will not only teach you the techniques but, also help you understand the psychological foundations behind them. This understanding is what will give you the edge—not just in marketing but in building lasting relationships with your audience.

I've always said that success leaves clues, and Andy has done an exceptional job of uncovering those clues in this book. Whether you're a seasoned marketer or someone just beginning to explore the field, the insights in *The Science of NeuroMarketing* will serve you well.

Take what you learn here, apply it, and watch your business—and your influence—grow.

Here's to your success.

Joseph McClendon III

Ultimate Performance Specialist
Speaker, Author, and Coach

THANK YOU

Thank you for embarking on this journey with me through the world of Neuromarketing. I am grateful that you chose to invest your time, energy, and trust into exploring the ideas and techniques shared within these pages.

Your commitment to growth and excellence in your marketing efforts is inspiring, and it has been my honor to guide you through these powerful concepts.

It is my sincere hope that the insights presented here will not only enhance your marketing skills but also empower you to connect more deeply with your audience, fostering relationships that are both meaningful and mutually beneficial.

I believe that marketing is more than just a tool for selling; it is a profound way to tell stories, build communities, and bring about positive change. Each chapter of this book was crafted with the intention of helping you unlock the full potential of your marketing efforts using NLP strategies, so you can make a genuine impact in the lives of your customers and clients.

As you integrate these strategies into your own work, I encourage you to remain curious, stay creative, and always strive for authenticity. Remember, the heart of great marketing lies in understanding and serving others. When you approach your work with empathy and integrity, you create not just successful campaigns, but lasting legacies.

I want to extend my heartfelt thanks to you, the reader, for your willingness to explore new ideas and push the boundaries of what is possible. Your journey does not end here—in fact, it is only just beginning.

I am excited for all the incredible things you will accomplish, and I am grateful to have been a part of your path toward marketing mastery.

Please know that your feedback, thoughts, and reflections are invaluable to me. I would love to hear how the concepts in this book have influenced your work and helped you grow. Feel free to reach out and share your experiences; together, we can continue to learn and evolve.

Thank you once again for your trust and for joining me on this adventure. Here's to your success, growth, and the exciting journey ahead!

With gratitude and best wishes,

Ai Andy

Andy "Ai Andy" Broadaway

Table of Contents

Introduction

In today's digital age, the internet has opened up unprecedented opportunities for businesses to reach a global audience. But with this vast potential comes intense competition. You're not just competing for market share—you're vying for attention in a world flooded with content, ads, and offers. The challenge isn't merely about being noticed; it's about standing out and compelling people to take action in a crowded marketplace.

You might have a revolutionary product or service. Still, without a message that resonates deeply with your audience, even the most innovative offerings can go unnoticed. Traditional marketing strategies often fail to tap into the deeper psychological triggers that influence human behavior. They focus on what you want to say rather than what your audience needs to hear.

This is where neuro marketing comes into play—a science-based approach that leverages the latest insights in psychology and neuroscience to craft marketing messages that don't just speak to your audience but drive them to act.

The Science Behind Neuromarketing

Neuromarketing is the intersection of marketing and neuroscience, focusing on how the brain responds to various

marketing stimuli. It's grounded in the understanding that purchasing decisions are not solely rational—they are deeply influenced by emotions, subconscious cues, and psychological triggers.

Studies have shown that up to 95% of purchasing decisions are made subconsciously, driven by emotions rather than logic. By understanding these processes, you can craft marketing strategies that resonate on a deeper, more instinctual level. Research in neuro marketing has uncover several key principles that can be used to influence consumer behavior:

- **Scarcity and Urgency**: Scarcity is a powerful motivator. When people perceive that a product is in limited supply, their desire for it increases. A study by Worchel, Lee, and Adewole (1975) found that cookies were rated as more desirable when they were scarce than when they were abundant. This principle taps into the brain's fear of missing out (FOMO), driving people to act quickly.
- **Social Proof**: Humans are inherently social creatures, and we often look to others for cues on how to behave. This is why testimonials, reviews, and endorsements can be incredibly effective. Research has shown that displaying social proof—such as "90% of people prefer this product"—can significantly increase conversion rates.
- **Emotional Triggers**: Emotions are a critical component of decision-making. Neuroscientist Antonio Damasio's research revealed that people with impaired emotional responses struggle to make even the simplest decisions. By tapping into emotions—whether through storytelling, imagery, or emotional language—you can create a more compelling narrative that resonates with your audience.

How Neuromarketing Can Grow Your Business

By integrating these psychological principles into your marketing strategies, you can transform how you connect with your audience. Neuromarketing isn't about manipulating people into making purchases; it's about understanding their needs and motivations on a deeper level and aligning your messaging accordingly.

1. **Increase Conversion Rates**: Neuromarketing techniques, such as using urgency or personalized messaging, can significantly boost conversion rates. For example, a study conducted by HubSpot found that personalized calls to action (CTAs) convert 202% better than generic ones. By understanding what drives your customers to take action, you can tailor your approach to increase engagement and conversions.
2. **Enhance Customer Loyalty**: When your marketing speaks directly to your audience's core values and desires, it fosters a deeper connection and builds trust. Trust is a critical component of customer loyalty. According to Edelman's Trust Barometer, 81% of customers say trust is a deciding factor in their buying decisions. Neuromarketing helps you build this trust by aligning your brand's messaging with your audience's psychological needs.
3. **Optimize Your Content Strategy**: Neuromarketing provides valuable insights into what types of content resonate most with your audience. By understanding which emotional triggers to tap into, you can create more effective content that not only attracts attention but also encourages sharing, boosting your brand's visibility and reach.

4. **Improve Advertising ROI**: Neuromarketing can also help you refine your advertising strategies to maximize return on investment (ROI). By focusing on what truly captures attention and drives action, you can create more compelling ads, reducing wasteful spending and improving overall campaign performance.

Real-World Applications of Neuromarketing

Many successful companies have already harnessed the power of neuromarketing to grow their businesses. For example, Coca-Cola's marketing campaigns often evoke feelings of happiness, nostalgia, and togetherness—emotional triggers that resonate with their target audience on a subconscious level. Apple's minimalist design and emphasis on social proof (like showcasing celebrities using their products) tap into desires for status and belonging.

Your Path to Business Growth

As an online entrepreneur, your goal is to turn attention into action and clicks into conversions. Neuromarketing equips you with the tools to achieve this by aligning your marketing strategies with human psychology. Instead of trying to out-shout the competition, you'll learn to speak the language of the mind—connecting with your audience on a profound, primal level.

By mastering these principles, you're not just learning to market better; you're learning to understand your customers better. This understanding will help you break through the noise, create messages that resonate, and ultimately drive growth for your business.

The science of neuromarketing is not just about boosting sales—it's about building lasting relationships with your customers. It's about creating experiences that are memorable, compelling, and, most importantly, meaningful.

Your journey into neuromarketing starts now, and with it, the potential to unlock new levels of success for your business.

References:

1. Zaltman, Gerald. *How Customers Think: Essential Insights into the Mind of the Market.* Harvard Business School Press, 2003.
2. Worchel, S., Lee, J., & Adewole, A. (1975). "Effects of Supply and Demand on Ratings of Object Value." *Journal of Personality and Social Psychology.*
3. Cialdini, Robert B. *Influence: The Psychology of Persuasion.* Harper Business, 2006.
4. Damasio, Antonio R. *Descartes' Error: Emotion, Reason, and the Human Brain.* Penguin Books, 1994.
5. HubSpot. "Personalized CTAs Convert 202% Better Than Default Versions." HubSpot Blog, 2016.

Edelman. "2020 Edelman Trust Barometer." Edelman, 2020

~CHAPTER I~

Anchoring:

Setting the Stage for Decision-Making

Every decision your customer makes is influenced by the information they receive first. This is where anchoring comes into play. Anchoring is a cognitive bias that occurs when an individual relies heavily on the first piece of information they encounter when making decisions.

In the context of online marketing, this initial information can significantly impact how your audience perceives the value of your products or services. When you present an anchor, it sets the stage for every subsequent judgment.

The Power of the First Impression

Research shows that people make up their minds about a brand in just 7 seconds, demonstrating the immense power of first impressions. The first price, feature, or benefit you mention becomes the reference point against which all other information is measured. This is why anchoring is so effective—it takes advantage of our natural tendency to rely on the initial piece of information as a baseline for all future decisions.

For example, if you're selling a high-ticket item, introducing the most expensive option first can make other, less expensive options seem more affordable by comparison. On the other hand, starting with a lower price point can anchor your audience's perception of what constitutes a good deal, making premium options feel out of reach.

Consider the weight loss program you're offering. If you initially present a $500 comprehensive package and then introduce a $200 basic package, the basic package will seem like a bargain in comparison.

The $500 anchor sets an expectation, and the lower price appears more attractive, even though $200 might have seemed expensive if introduced first. This demonstrates how anchoring can shape perceptions and guide decisions based on initial information.

Anchoring Beyond Pricing

The effectiveness of anchoring doesn't just stop at pricing. It also plays a crucial role in how you present features and benefits. Let's say you're highlighting the benefits of some survival gear. If you start by emphasizing the life-saving aspects of the gear—such as its ability to purify water or start a fire in the wild—everything else you mention afterward, like durability or weight, will be viewed through the lens of these primary, life-critical benefits. The initial anchor—the life-saving potential—makes other features seem like added bonuses, rather than primary selling points.

Anchoring is also useful in guiding your audience's decision-making process when offering multiple options. Imagine you're selling a series of online marketing courses. If you first introduce an all-inclusive package with extensive content and personal coaching for $1,000, this becomes the anchor.

When you follow it with a more limited course for $500, your audience is likely to perceive the second option as a reasonable alternative, even if they initially balked at the $1,000 price tag.

The anchor of $1,000 makes $500 seem more accessible, and for those who decide to invest in the higher-priced option, it solidifies their commitment by making them feel they are receiving premium value.

Leveraging Anchoring for Maximum Impact

But anchoring isn't just about presenting a higher price first. The strategy can work in the opposite direction as well. If your goal is to steer customers toward a premium option, you might start with a very basic, low-cost anchor that makes the higher-priced options seem more valuable.

For instance, if you first offer a free trial or a $10 introductory course, the next tiered offer at $150 feels more substantial, packed with more value. The initial low-cost anchor sets an expectation that's easily exceeded by your premium offerings. Beyond pricing and features, anchoring can influence how customers perceive time and effort.

Suppose you're offering a 12-week weight loss plan. By first discussing the challenges of losing weight quickly and unsustainably, you anchor the idea that lasting weight loss requires time and commitment. When you then present your 12-week plan, it seems more realistic and appealing because it's positioned as the healthier, more achievable alternative to the unrealistic quick fixes that often fail.

Real-World Applications of Anchoring

Anchoring also has a significant impact on how you frame customer testimonials and case studies. When you highlight a particularly dramatic success story first, such as a client who achieved outstanding results in a short time, you set an anchor that raises expectations for what your product or service can deliver.

Subsequent testimonials, even if they reflect more modest successes, will be viewed in light of that initial success story, making them seem more impressive by association. In an online environment, where your audience is bombarded with information, anchoring helps you control the narrative.

It allows you to guide your customers toward the conclusions you want them to reach without overtly telling them what to think. This subtle influence is often more effective than direct persuasion because it feels organic and self-directed.

For example, when crafting your sales page, consider the sequence in which you present information. Start with the most compelling, value-driven aspect of your offer—whether it's a high price point, a key benefit, or a powerful testimonial. This initial anchor will shape how your audience perceives everything else that follows.

If you bury your most important point in the middle or end of your content, you lose the opportunity to set a strong anchor that can guide decision-making. Anchoring can also be integrated into your email marketing strategy. When you send out a campaign, the first subject line your audience sees can anchor their expectations for the content inside.

If you lead with a subject line that emphasizes exclusivity or high value, the content of your email will be perceived through that lens, making your offers and calls to action more persuasive.

Anchoring in Content and Ethical Considerations

Even in content marketing, anchoring plays a role. The first paragraph of your blog post or article sets the tone for everything that follows. If you start with a bold statement or an intriguing fact, you anchor your readers' expectations and keep them engaged, guiding them to view the rest of the content as valuable and worth their time.

However, anchoring must be used strategically and ethically. Over-reliance on this technique, or using it in a way that feels manipulative, can backfire. Your audience is smart, and they value transparency.

If they sense that you're trying to manipulate their perceptions without providing real value, they may lose trust in your brand. The key is to use anchoring to highlight genuine value, not to obscure it.

Conclusion: Setting the Stage for Success

Ultimately, anchoring is about setting the stage for how your audience interprets the information you present. It's a powerful tool that, when used correctly, can significantly influence customer behavior. By understanding and applying anchoring in your marketing strategies, you can guide your audience toward the decisions that benefit both them and your business.

As Tony Robbins once said, *"It's not about the goal. It's about growing to become the person that can accomplish that goal."* In the same vein, anchoring is about shaping the customer journey to align with both your goals and theirs, fostering a path to mutual growth and success. Whether you're pricing products, presenting features, or sharing testimonials, the way you frame that first piece of information will shape everything that follows, leading to more effective and profitable outcomes.

By mastering the art of anchoring, you set the foundation for success, turning initial impressions into long-lasting customer relationships. Remember, it's not just about influencing decisions—it's about understanding the psychology behind them and using that insight to create genuine value.

Conclusion: Setting the Stage for Success

Anchoring is a powerful cognitive tool that can shape how your audience perceives and interacts with your brand. By strategically presenting initial information—whether it's pricing, features, benefits, or even customer testimonials—you can guide your customers' decision-making processes in subtle yet impactful ways.

This technique is about much more than manipulation; it's about setting the stage for meaningful, value-driven engagements that foster trust and loyalty.

When used ethically, anchoring helps you communicate the true value of your offerings and align your customer's expectations with your business goals. It creates a foundation upon which all subsequent interactions are built, leading to stronger customer relationships and more profitable outcomes.

As the famous quote by Robert Collier goes, *"Always enter the conversation already taking place in the customer's mind."* Anchoring allows you to do just that—joining your audience where they are and guiding them toward where you want them to be. By mastering anchoring, you're not just setting up a sale; you're setting up a journey. A journey that benefits both you and your customers, fostering a path to mutual growth and success. Whether through pricing strategies, the presentation of features, or how you share testimonials, anchoring can be the difference between a passing interest and a loyal customer.

Call to Action

Ready to harness the power of anchoring in your marketing strategy?

Start by examining how you currently present your offers. Are you setting the right anchors to guide your customers' perceptions and decisions? Identify key points in your marketing where a strategic anchor could enhance customer engagement and conversions.

Take action today—refine your messaging, adjust your pricing strategies, and optimize your content to set strong, value-driven anchors. By doing so, you'll not only increase conversions but also build lasting relationships based on trust and value.

The power to influence and persuade is in your hands—use it wisely and see your business thrive.

~CHAPTER 2~

Scarcity

Creating Urgency Through Limited Availability

Scarcity taps into a fundamental human instinct: the fear of missing out (FOMO). This fear is so ingrained in us that when we perceive something as limited or fleeting, its value in our eye's skyrockets. This principle is deeply rooted in our psychology, driving us to take action when we feel an opportunity might slip away. For you as an online entrepreneur, leveraging scarcity can be one of the most effective strategies to motivate your audience to act quickly.

Scarcity operates on a basic psychological tenet: people assign more value to things that are less available. Research shows that scarcity can significantly increase perceived value and desirability. A study by Worchel, Lee, and Adewole (1975) demonstrated that participants rated cookies as more desirable when there were only two cookies in a jar compared to when there were ten.

If your product or service is perceived as rare or time-sensitive, your audience is more likely to want it. This isn't just about creating artificial limitations; it's about highlighting the genuine value and urgency of what you offer.

The Psychological Impact of Scarcity

Imagine you're running a weight loss program and have only a limited number of spots available. By clearly communicating that only a few places remain, you create a sense of urgency. Potential customers who might have otherwise hesitated are more likely to sign up immediately, fearing that they'll miss out if they delay. The limited availability of your program makes it more appealing, as people instinctively want what they think they can't easily have.

Scarcity is effective because it triggers a psychological response that leads to quicker decision-making. When people perceive a resource as scarce, their brain goes into a "scarcity mindset," heightening their focus and making them prioritize that resource. This is why limited-time offers and exclusive deals work so well—they capture the audience's attention and compel them to act fast.

Scarcity in Product Launches

The power of scarcity is also evident in product launches. When you release a new product and announce that it's available for a limited time or in limited quantities, you're tapping into this psychological trigger. A study by the Journal of Consumer Research found that products perceived as scarce due to limited quantity or time are viewed as more valuable, prompting quicker buying decisions. Consumers are more likely to purchase if they believe the opportunity won't last.

This technique can be especially effective when combined with real-time updates, such as showing how many items are left in stock or how many spots are available in a course. These tactics reinforce the perception of scarcity, driving customers to act before it's too late.

Digital Scarcity: Creating Urgency with Intangible Products

But scarcity isn't just about physical products or services. It can also be applied to digital content. If you're offering an exclusive webinar or a downloadable guide for a limited time, you're leveraging scarcity to increase participation. The fact that the content won't always be available adds value and motivates your audience to engage now rather than later.

This is particularly powerful in the online marketing world, where digital products can often feel less tangible and, therefore, less urgent. By applying the principles of scarcity, you can elevate the perceived value of your digital offerings, prompting quicker action from your audience.

Scarcity in Pricing and Exclusivity

Scarcity can also be effectively used in pricing strategies. Early-bird discounts, for example, create a window of opportunity where customers can get a better deal if they act fast. Once that window closes, the higher price reinforces the idea that those who acted quickly were smart to do so. Research by Behavioral Economist Dan Ariely indicates that scarcity can enhance consumer decision-making and satisfaction by providing a clear incentive to act without overthinking. This doesn't just incentivize early purchases; it also reinforces the value of your product by showing that it's worth more as time goes on.

Another way to implement scarcity is through exclusive offers. If you have a segment of your audience that is particularly engaged, offering them something that's not available to the general public can create a sense of exclusivity and urgency.

For example, if you're in the survival niche, you might offer your most loyal customers early access to a new product before it's released to everyone else. This not only rewards your best customers but also creates a buzz that can lead to even greater demand when the product is launched widely.

Timing and Credibility in Scarcity Marketing

Scarcity can also be tied to timing. Limited time offers are a classic example of this. By setting a deadline for a special deal or discount, you push your audience to make a decision quickly.

If they know that the offer will expire soon, they're less likely to procrastinate. This strategy is particularly effective during holiday seasons or product launches, where timing is everything, and consumers are already in a buying mindset.

However, for scarcity to be effective, it needs to be credible. If your audience senses that the limitations you're presenting are artificial or manipulative, they might lose trust in your brand. This is why it's essential to be honest about the scarcity you're creating. If you only have a limited number of spots in your program because of capacity constraints, say so. If your discount is truly limited-time, make sure it doesn't reappear a week later. Authenticity in scarcity not only drives urgency but also builds trust.

The Effective Use of Scarcity in Building Trust and Urgency

In online marketing, where customers are bombarded with choices, scarcity helps you stand out. It gives your audience a reason to choose you now rather than later, cutting through the noise of endless options. By creating genuine urgency, you guide your customers toward making quicker decisions, which can lead to higher conversion rates and more immediate sales.

The effectiveness of scarcity lies in its ability to create a sense of urgency that moves people to act. As marketing expert Seth Godin once said, *"People don't buy goods and services. They buy relations, stories, and magic." Scarcity creates that "magic"* moment where customers feel compelled to act, fearing they might miss out on something valuable.

When your audience feels that they might miss out on something valuable, they're far more likely to take the leap. This doesn't just increase sales; it enhances the perceived value of your offerings, making your customers feel they've made a smart, timely decision.

By leveraging scarcity effectively, you create a win-win scenario where both you and your customers benefit—your business sees increased sales and engagement, while your customers feel satisfied with their choice, knowing they acted swiftly to seize a valuable opportunity.

Conclusion:

Scarcity is a potent psychological trigger that taps into the fundamental human fear of missing out. By creating a sense of urgency and highlighting the limited availability of your products or services, you can drive quicker decision-making and increase conversions. Whether through limited-time offers, exclusive deals, or emphasizing the rarity of a digital product, scarcity can make your audience feel a heightened desire to act immediately.

However, it's crucial to use scarcity authentically and ethically to build trust and credibility with your audience. When applied thoughtfully, scarcity not only enhances the perceived value of your offerings but also fosters stronger customer relationships by making them feel like they're part of something special and exclusive.

Call to Action

Ready to implement scarcity in your marketing strategy?

Start by identifying a product or service where you can create a genuine sense of urgency. Whether it's a limited-time offer, an exclusive launch, or highlighting limited availability, use these techniques to drive action. Experiment with different approaches and see how your audience responds.

Don't miss out on the opportunity to boost your sales and engagement—leverage the power of scarcity today to make your offerings more compelling and irresistible.

~CHAPTER 3~

Social Proof

Leveraging the Power of Group Influence

Social proof is a powerful psychological phenomenon that plays a critical role in decision-making. When people are uncertain about what action to take, they look to the behavior of others for guidance. This tendency, rooted in our evolutionary past, is driven by the desire to make the "correct" decision when faced with ambiguity. As social creatures, we have evolved to trust the judgment of others, especially when we perceive them as being similar to ourselves.

In online marketing, this principle can be a game-changer. By showcasing the actions, endorsements, and experiences of others, you can influence potential customers to follow suit, making them more likely to engage with your brand, buy your products, or sign up for your services.

The Psychology Behind Social Proof

Social proof taps into the "herd mentality," a concept that describes how people tend to conform to the actions of a larger group. This psychological principle is so influential that studies have shown people are more likely to take action if they see others doing the same.

For instance, research by Dr. Robert Cialdini, a pioneer in the field of persuasion, demonstrated that people are significantly more likely to reuse towels in a hotel when informed that the majority of other guests had done so. This effect occurs because people naturally assume that if others are doing something, it must be the right thing to do.

As Warren Buffett once said, *"It takes 20 years to build a reputation and five minutes to ruin it."* Social proof helps build that reputation by demonstrating that others have already vetted and approved of your brand.

Testimonials and Trust

Imagine you're selling a survival course. If you display testimonials from people who have successfully used your course to prepare for emergencies, potential customers are more likely to trust your expertise. The experiences of others serve as evidence that your course is effective, reducing any doubts or hesitations they might have. When they see that others like them have benefited, they're more inclined to believe that they will too.

This aligns with the concept that *"people like us do things like this,"* a phrase coined by marketing expert Seth Godin to describe how individuals are influenced by their peers' actions. Social proof isn't limited to testimonials. It can take many forms, each of which can be strategically used depending on your audience and goals.

For example, displaying the number of customers who have purchased your product, the number of followers on your social media accounts, or the ratings and reviews on your website all serve to reinforce the idea that your product or service is popular and trusted by others. A Nielsen study found that 92% of people trust recommendations from friends and family over any other type of advertising.

This demonstrates the immense power of social proof, particularly when it comes from credible sources that are relatable to your target audience.

User-Generated Content: The New Word of Mouth

Another powerful form of social proof is user-generated content. When your customers share photos, videos, or stories about your product, they're not just providing testimonials—they're creating authentic, relatable content that resonates with potential buyers. User-generated content is particularly persuasive because it comes from real people who are not directly affiliated with your brand, making it seem more genuine.

A study by Turn To Networks found that 90% of consumers say user-generated content influences their purchase decisions more than other types of advertising. For instance, if your weight loss supplement has inspired customers to share their before-and-after photos, showcasing these images on your site or social media can significantly boost your credibility and influence new customers to try the product.

Influencer Marketing: Leveraging Credibility

Social proof can also be harnessed through influencer marketing. When a respected figure in your niche endorses your product, their followers are more likely to view it as credible and worth purchasing. This is because influencers have built a relationship of trust with their audience, and their endorsement serves as a powerful form of validation.

If a well-known survival expert recommends your gear, for example, their followers are likely to see your product as reliable and high-quality. According to a survey by Edelman, 63% of consumers trust influencers' opinions about products "much more" than what brands say about themselves. This shows that influencer endorsements can be an effective way to build trust and sway potential customers.

Awards, Certifications, and Recognitions

In addition to testimonials and endorsements, social proof can be created by showcasing awards, certifications, or other forms of recognition. If your product has been endorsed by a reputable organization, this adds another layer of credibility. Potential customers who might be on the fence can be swayed by the fact that your product has been recognized by experts in the field. Research shows that displaying certifications or awards can increase conversion rates by up to 30%, as they provide assurance to customers about the quality and reliability of your offerings.

Social Media and the Power of Engagement

Social proof also extends to social media. The number of likes, shares, and comments on your posts can influence how others perceive your brand. When people see that your content is popular and engaging, they're more likely to interact with it themselves. This is why it's important to cultivate an active and engaged social media presence—every like or share serves as a mini endorsement that can help attract new customers. The mere presence of social proof, like a large number of followers or high engagement rates, can increase perceived credibility and encourage further engagement.

Authenticity and Relevance in Social Proof

However, the most effective social proof is not just about quantity but also relevance. It's crucial to showcase testimonials, user-generated content, and endorsements that resonate with your target audience.

If you're marketing to fitness enthusiasts, for example, testimonials from other fitness enthusiasts will be more compelling than those from people in unrelated fields. The more your potential customers can relate to the experiences of others, the more likely they are to be influenced by them.

It's also important to present social proof in a way that feels authentic and unforced. Overloading your website with too many testimonials or pushing influencer endorsements too hard can backfire, making your efforts seem insincere. Instead, aim for a balanced approach where social proof is integrated naturally into your content and marketing materials. Highlight the most compelling and relevant examples without overwhelming your audience.

Building a Community of Trust

Social proof works best when it's a natural extension of your brand's story. When people see others engaging with your product in a genuine way, they're more likely to feel confident about doing the same. As Brene Brown, a renowned researcher on vulnerability and trust, once said, *"Trust is built in very small moments."* Every piece of social proof is one of those moments, gradually building a narrative that potential customers can trust.

Whether it's through testimonials, user-generated content, influencer endorsements, or social media engagement, the goal is to create a sense of community around your brand, where potential customers feel like they're joining a group of satisfied users.

Conclusion:

In a competitive market, social proof can be the deciding factor that sets you apart. It's not just about what you say about your product; it's about what others say. By leveraging the power of group influence, you can build trust, reduce uncertainty, and motivate potential customers to take action.

When people see that others are already on board, they'll want to be part of it too. As Zig Ziglar, a famous motivational speaker, once said, *"People don't buy for logical reasons. They buy for emotional reasons."* Social proof taps into those emotional reasons by showing that others have already made a similar choice and are happy with it, which in turn, makes the choice feel safer and more compelling.

Use social proof effectively, and you create not just a marketing tool, but a powerful community-driven strategy that brings people together around your brand.

Call to Action

Ready to harness the power of social proof in your marketing strategy? Start by gathering and showcasing testimonials from satisfied customers, encouraging user-generated content, or partnering with influencers in your niche. Implement these strategies today to build trust, increase engagement, and drive conversions. Don't just tell your audience why they should choose your brand—show them why others already have. Begin leveraging social proof now and watch your brand's credibility and customer base grow!

~CHAPTER 4~

Reciprocity

The Science of Give & Take in Marketing

Reciprocity is a cornerstone of human interaction. It's the idea that when someone does something for you, you feel compelled to return the favor. This principle, deeply ingrained in our social fabric, can be a powerful tool in your marketing strategy. By giving something valuable to your audience, you can encourage them to reciprocate, often by making a purchase, signing up for a service, or engaging with your brand in a meaningful way.

The Psychology of Reciprocity

At its core, reciprocity in marketing is about creating a sense of obligation. When you provide value to your audience without asking for anything in return, they naturally feel a need to give back. This is not about guilt; rather, it's about building a positive relationship based on mutual benefit. Dr. Robert Cialdini, in his book *Influence: The Psychology of Persuasion*, describes reciprocity as one of the most potent tools of influence, stating, "People say yes to those they owe."

The key to leveraging this principle effectively is ensuring that what you offer is genuinely valuable and not just a superficial gesture. Consider offering a free eBook, guide, or resource in the weight loss niche. By providing detailed, actionable content that helps your audience solve a problem or achieve a goal, you're giving them something of real worth.

They didn't pay for it, yet they received value, and now they're-more-inclined. To explore what else you offer. This could lead them to purchase your premium products or services because they already trust that you deliver value.

Free Trials and the Reciprocity Effect

The concept of reciprocity also applies to free trials or samples. By allowing potential customers to experience your product at no cost, you're not only showcasing its value but also triggering the principle of reciprocity. Studies have shown that free trials can increase conversion rates by up to 30% because people often feel a sense of loyalty or obligation to continue the relationship after receiving something for free.

When people receive something for free, they often feel compelled to reciprocate, which can lead to conversions once the trial period ends. This is particularly effective in subscription-based services or software platforms, where the initial free offering demonstrates the utility and benefits of the product, making the transition to a paid version seem like a natural next step.

Exclusive Content and Early Access

Another way to leverage reciprocity is through exclusive content or early access. If you run a survival blog, for example, you might offer your most dedicated followers' early access to a new course or a special report. By making them feel like they're part of an inner circle, you're not only providing value but also strengthening their connection to your brand.

In return, they're more likely to support your future endeavors, whether that's by purchasing additional products, sharing your content, or even becoming advocates for your brand.

Building Relationships Through Genuine Engagement

Reciprocity is also at play in the way you interact with your audience. Offering personalized advice, answering questions, or providing exceptional customer service are all forms of giving that can inspire loyalty and lead to increased sales.

Research indicates that 73% of consumers say friendly customer service representatives can make them fall in love with a brand. When people feel that you genuinely care about their needs and are willing to help them without immediately asking for something in return, they're more likely to support your business.

This approach can be particularly powerful in niches where trust and expertise are crucial, such as health and wellness or financial services. In the words of Zig Ziglar, *"You will get all you want in life, if you help enough other people get what they want."*

Email Marketing and Reciprocity

In email marketing, reciprocity can be used to build stronger relationships with your subscribers. Offering free resources, tips, or exclusive deals in your emails creates a sense of goodwill, making your audience more likely to open future emails and engage with your content.

Over time, this builds trust and increases the likelihood that they'll take advantage of your paid offerings. The key is to ensure that your free content is valuable enough to make them feel they've received something of worth, which in turn makes them more willing to reciprocate by becoming a paying customer.

The Power of Small Gestures

Reciprocity isn't just about large gestures, either. Small acts of kindness, like thanking customers for their business or acknowledging their feedback, can also create a positive cycle of give and take. Studies have shown that expressing gratitude can increase customer satisfaction and loyalty, with 86% of customers reporting that they are more likely to continue doing business with companies that express gratitude.

These gestures, while seemingly minor, can have a significant impact on customer loyalty. When people feel appreciated and valued, they're more likely to return the favor by continuing to do business with you, leaving positive reviews, or recommending your brand to others.

Leveraging Reciprocity in Partnerships

The power of reciprocity also extends to partnerships and collaborations. By promoting another brand's products or sharing their content with your audience, you're not only providing value to your followers but also fostering goodwill with that brand.

This can lead to reciprocal promotions, joint ventures, or other mutually beneficial opportunities that help both businesses grow. In the competitive world of online marketing, building these kinds of relationships can be a key factor in your long-term success.

The Importance of Authenticity

However, it's important to approach reciprocity with sincerity. If your audience senses that your "gifts" are merely a ploy to manipulate them into buying, they may become distrustful and disengaged. The value you provide must be genuine and tailored to their needs.

When done correctly, reciprocity doesn't feel like a transaction—it feels like a natural part of a relationship. Your audience gives back because they want to, not because they feel forced. As Maya Angelou wisely stated, *"People will forget what you said, people will forget what you did, but people will never forget how you made them feel."*

Conclusion

Reciprocity works because it taps into a fundamental aspect of human nature: the desire to repay kindness. By giving freely and genuinely, you create a positive cycle that benefits both you and your audience. They receive value, and in return, they're more likely to support your business, whether through purchases, engagement, or word-of-mouth recommendations.

In a world where customers are constantly bombarded with sales pitches, reciprocity offers a refreshing and effective way to build trust and foster lasting relationships.

Call to Action (CTA)

Now it's your turn to put the principle of reciprocity into action. Start by giving something valuable to your audience—whether it's a free guide, exclusive content, or simply exceptional customer service. Watch as this simple act of giving transforms your relationship with your audience and drives your business growth.

Ready to make a difference?

Get started today by offering something valuable and see the power of reciprocity at work!

~CHAPTER 5~

Color Psychology

Using Colors to Influence Customer Behavior

Color psychology is a subtle yet powerful tool in marketing. The colors you choose for your branding, website, and marketing materials can significantly influence how your audience perceives your business and how they behave. Colors evoke emotions and reactions that often operate below the level of conscious awareness, making them a critical component in shaping customer behavior. Every color carries its own set of associations and emotional triggers.

The Impact of Color on Customer Perception

Research has shown that up to 90% of snap judgments made about products can be based on color alone, depending on the product. This highlights the importance of selecting the right colors to create the desired impact. For example, red is often associated with urgency, passion, and excitement. It's a color that grabs attention and can increase the heart rate, making it a popular choice for calls to action or clearance sales. When you want your audience to act quickly, red can be the push they need.

But red isn't just for urgency—it's also a color that can signify power and confidence, which can be useful in building a bold, assertive brand identity. Blue, on the other hand, conveys trust, calmness, and reliability. It's no coincidence that many financial institutions and tech companies use blue in their logos and branding.

Studies have found that blue can increase feelings of trust and security, which is why it is often used in industries where credibility is paramount. In the weight loss niche, for instance, using blue can help reinforce a message of stability and long-term commitment to health, making your brand feel like a trusted partner in your customers' journey.

Green, Yellow, and Their Psychological Associations

Green is another color rich in psychological implications. It's often associated with health, growth, and tranquility. In the survival niche, green can evoke the natural world and the idea of self-sufficiency, making it an excellent choice for products or services aimed at outdoor enthusiasts or those interested in sustainable living.

According to a study by the University of Rochester, the color green can enhance creative thinking, which can be a valuable insight for brands looking to foster innovation and sustainability. Yellow is a color of optimism and energy. It's bright, cheerful, and can draw attention without being as aggressive as red. If your goal is to create a sense of happiness or to evoke a sunny, positive vibe, yellow can be highly effective.

However, it's a color that should be used carefully, as too much yellow can sometimes feel overwhelming or can create visual strain.

In online marketing, yellow works well in moderation, particularly when you want to highlight key features or benefits that make your audience feel good about their choices.

Black, White, Orange, and Purple: Strategic Color Choices

Black and white, while technically not colors, play crucial roles in design and branding. Black is associated with luxury, sophistication, and power. It's a color that can make a brand feel exclusive and high-end.

If you're selling premium products, using black in your branding can create a sense of elegance and desirability.

White, in contrast, represents simplicity, cleanliness, and purity. It's often used to create a minimalist aesthetic that allows other colors to stand out, making it a perfect backdrop for highlighting vibrant colors or for creating a modern, clean look. Orange is a color of enthusiasm and creativity. It combines the energy of red with the cheerfulness of yellow, making it a great choice for brands that want to convey fun and excitement.

Orange can be particularly effective in industries like fitness, where you want to inspire action and a positive, energetic attitude. It's a color that encourages engagement and can make your brand feel approachable and lively. Research has shown that orange can increase appetite, making it a popular choice for food brands.

Purple is often associated with luxury, mystery, and spirituality. It's a color that can evoke a sense of depth and wisdom, making it a good choice for brands that want to convey a sense of sophistication or intrigue.

In the wellness or self-improvement niches, purple can suggest a connection to higher ideals or personal transformation, appealing to customers who are looking for something beyond the ordinary.

Combining Colors for Maximum Impact

The psychology of color isn't just about the individual meanings of each hue. It's also about how colors work together. Complementary colors, for example, create contrast and can make certain elements pop, drawing the eye to where you want it. Analogous colors, which are next to each other on the color wheel, create harmony and a more subtle, pleasing effect.

The way you combine colors can significantly impact the mood and tone of your marketing materials, guiding how your audience feels as they interact with your brand. In online marketing, the choice of colors can influence everything from the mood of your website to the effectiveness of your call-to-action buttons. A red button might drive more conversions because it stands out and creates a sense of urgency, while a blue button might convey trustworthiness, making customers feel secure as they complete a purchase.

Even the background color of your website or landing pages can impact user experience and engagement. A light, neutral background can make content easier to read and keep the focus on your key messages, while a dark background can create a dramatic, immersive experience.

The Role of Color in Branding and Cultural Sensitivity

Color psychology also extends to your branding and logo design. The colors you choose for your logo will be the first impression customers have of your brand, setting the tone for their entire experience.

A well-chosen color palette can make your brand memorable and distinctive, helping you stand out in a crowded marketplace. Consistency in using your brand colors across all touchpoints—your website, social media, email marketing, and packaging—reinforces your brand identity and makes it easier for customers to recognize and trust your brand.

However, the impact of color can vary depending on your audience. Cultural differences, personal preferences, and even the context in which the color is seen can all influence how it's perceived. For example, while white is often associated with purity in Western cultures, it can signify mourning in some Eastern cultures. This is why it's essential to test different color combinations and gather feedback from your audience to see what resonates best.

What works for one niche or demographic might not work for another, so being attuned to the nuances of color perception is key to using it effectively.

Conclusion

Color psychology is a subtle yet powerful tool that, when used thoughtfully, can shape how your audience perceives and interacts with your brand. It's not just about making things look good; it's about creating the right emotional response that aligns with your marketing goals. As Pablo Picasso once said, *"Colors, like features, follow the changes of the emotions."* Whether you're trying to build trust, create urgency, or evoke a sense of luxury, the colors you choose play a crucial role in guiding customer behavior.

By understanding and applying the principles of color psychology, you can enhance your marketing efforts and create a more engaging, effective brand experience.

Call to Action (CTA)

Ready to transform your brand with the right colors?

Take a closer look at your marketing materials today and consider how a change in color strategy could boost your conversions and customer engagement. Experiment, test, and watch your brand come to life with the power of colors.

~CHAPTER 6~

Flattery

The Subtle Art of Winning Over Customers

Flattery is a subtle but effective way to win over customers. It involves complimenting your audience in a way that makes them feel good about themselves, their choices, and their potential decisions. When used wisely, flattery can build a positive emotional connection between your brand and your customers, making them more likely to engage with your content, trust your recommendations, and ultimately make a purchase.

The Psychology of Flattery

Flattery works because it taps into a basic human need: the desire to feel valued and appreciated. Research in social psychology shows that when individuals receive compliments, it activates the same reward centers in the brain as receiving a tangible reward. This suggests that flattery can be as influential as more direct incentives in motivating behavior.

When you flatter your audience, you're not just giving them a compliment; you're validating their identity and affirming their choices. This creates a sense of rapport and makes them more receptive to your message. However, the key to effective flattery is sincerity. Customers can quickly detect insincerity, and if they feel they're being manipulated, the tactic can backfire. A study published in the *Journal of Personality and Social Psychology* found that insincere flattery can lead to negative feelings and distrust. To avoid this, your compliments need to be genuine and relevant to your audience's needs and values.

Personalized Marketing: Flattery in Action

Consider the impact of a personalized email campaign. If you're marketing a weight loss program, an email that begins with a statement like, "*We know how committed you are to achieving your health goals, and we're here to help you every step of the way,*" flatters the recipient by acknowledging their dedication. This type of compliment makes the recipient feel seen and appreciated, increasing the likelihood that they'll continue reading and consider your offer.

By recognizing their efforts and aspirations, you create a connection that goes beyond a simple sales pitch. According to a study by Experian, personalized emails deliver six times higher transaction rates, demonstrating the power of tailored messaging and the role of flattery in enhancing engagement.

Flattery can also be woven into your product descriptions. For example, if you're selling survival gear, you might describe the customer as someone who is smart and prepared, someone who takes safety seriously and is always ready for the unexpected. This subtle flattery not only boosts their self-image but also reinforces the idea that purchasing your product aligns with their self-perception as a responsible and capable individual.

The result is a stronger emotional bond with your brand, which can lead to higher sales and greater customer loyalty.

Social Media and Customer Engagement

Another area where flattery can be effective is in your social media interactions. Responding to comments or messages with positive reinforcement, such as, "You have great taste—this is one of our most popular products!" can make your followers feel good about their choices.

This kind of flattery can turn casual followers into loyal fans who are more likely to engage with your content, share your posts, and recommend your products to others.

It's about creating a community where your audience feels valued and recognized. Social media is not just a platform for marketing; it's a space for building relationships, and flattery can be an effective way to deepen those relationships.

Flattery in Customer Service and Feedback

Flattery can also be used in customer service. When dealing with a customer inquiry or complaint, acknowledging the customer's insight or patience can go a long way. Phrases like, "You're absolutely right, that's a great observation," or "We really appreciate your understanding and feedback," flatter the customer by making them feel knowledgeable and valued. This can turn a potentially negative interaction into a positive one, reinforcing their loyalty to your brand and making them more likely to continue doing business with you.

However, flattery should be used with care, especially in marketing copy. Overdoing it can come off as insincere or even patronizing, which can erode trust. The best approach is to use flattery sparingly and in a way that feels natural and appropriate for the situation.

For example, instead of constantly telling your customers how amazing they are, focus on specific aspects that genuinely reflect their actions or choices.

Testimonials, Reviews, and Flattery

In online reviews and testimonials, flattery can also play a role. Highlighting customer quotes that praise the product while subtly complimenting the customer's choice can be an effective way to use flattery without being overt. For instance, a review that says, *"I'm so glad I chose this program—it's really helped me achieve my goals,"* not only praises the product but also affirms the customer's decision-making skills. Sharing these types of reviews on your website or social media not only provides social proof but also subtly flatters potential customers by suggesting that they, too, can make a smart choice by choosing your product.

Flattery is also effective in upselling and cross-selling. When you suggest additional products, you can flatter the customer by positioning the suggestion as something that would enhance their already excellent choice. For example, "Since you've chosen our premium package, which is a fantastic decision, we recommend adding this accessory to get even more out of your purchase." This approach flatters the customer's initial choice while also encouraging them to consider additional purchases, increasing your overall sales.

The Emotional Power of Flattery

The success of flattery in marketing lies in its ability to create positive emotions. When people feel good about themselves, they're more likely to associate those positive feelings with your brand. This emotional connection can be a powerful motivator, driving customers to choose your products over competitors' and fostering long-term loyalty.

The key is to keep it genuine, relevant, and appropriate for the context, ensuring that your audience feels appreciated rather than manipulated. As Oprah Winfrey once said, *"People want to be heard. They want to be validated."* Flattery is a way of hearing and validating your customers, recognizing and affirming the qualities and decisions that make them special.

When done right, it can enhance your relationship with your audience, making them feel valued and understood. This not only increases their likelihood of making a purchase but also strengthens their connection to your brand, turning them into loyal customers and advocates who are more likely to return time and time again.

Conclusion

Flattery is not about empty praise; it's about creating genuine connections by recognizing and affirming your customers' qualities and decisions. When applied thoughtfully, flattery can enhance your marketing strategy, increase customer engagement, and foster loyalty.

Call to Action (CTA)

Now it's your turn to put the art of flattery into practice. Start by incorporating sincere compliments into your marketing communications—whether it's through personalized emails, social media interactions, or customer service. Watch how this subtle yet powerful tool can transform your customer relationships and drive your business growth. Ready to make your customers feel valued?

Begin today by mastering the art of flattery in your marketing!

~CHAPTER 7~

Storytelling

Crafting Narratives That Resonate and Convert

Storytelling is one of the most powerful tools in marketing. It goes beyond simply presenting facts or listing benefits. Instead, it taps into the human desire for connection, emotion, and meaning. A well-crafted story can make your brand memorable, build a deeper relationship with your audience, and drive conversions by resonating on a personal level. When you tell a story, you're not just selling a product or service; you're offering an experience, an identity, and a solution wrapped in a narrative that your customers can relate to.

The Emotional Power of Storytelling

At the heart of effective storytelling is the ability to connect with your audience's emotions. Research shows that people make decisions based on emotions and then justify them with logic. A study by neuroscientist Paul Zak found that character-driven stories with emotional content can increase levels of oxytocin, a hormone associated with empathy, leading people to feel more connected and motivated to take action. By telling a story that speaks to your audience's fears, hopes, or dreams, you can create an emotional connection that makes your message more impactful.

For example, if you're marketing a weight loss program, rather than just listing its benefits, you could share a story about someone who struggled with their weight for years, tried multiple programs without success and finally found your program, which changed their life.

This kind of narrative makes the potential customer see themselves in the story and feel that your product could be the solution to their struggles. As author and marketing expert Donald Miller says, *"People don't buy the best products; they buy the products they can understand the fastest."* A compelling story makes your product easy to understand and relate to.

Humanizing Your Brand

A great story also humanizes your brand. It shows that there are real people behind the products and services, people who understand the challenges and desires of their customers. By sharing the story of how your company started, the obstacles you faced, and the mission that drives you, you invite your audience to be part of your journey. This transparency builds trust and loyalty, as customers feel they are supporting a brand that aligns with their values and experiences.

Simon Sinek, in his book *Start with Why*, emphasizes that "*People don't buy what you do; they buy why you do it.*" Storytelling allows you to communicate your "why" in a way that resonates with your audience, making them more likely to support your brand.

Turning Features into Benefits

Storytelling is particularly effective in turning features into benefits. While a feature might tell what a product does, a story shows how that feature improves someone's life. In the survival niche, for instance, instead of just stating that the gear is durable and reliable, you could tell the story of a customer who used your gear during a challenging outdoor adventure and how it performed under extreme conditions, ensuring their safety.

This narrative not only demonstrates the product's value but also places the customer in the story, imagining themselves in similar situations where the product could be a lifesaver.

According to a study by the *Harvard Business Review*, people are more likely to remember information when it is presented in a story format than when it is presented in a list of facts or figures. This means that by telling stories, you can make your product's benefits more memorable and persuasive.

Creating Identity and Belonging

Another powerful aspect of storytelling is its ability to create a sense of identity and belonging. People want to be part of something bigger than themselves, to belong to a group that shares their values and goals. By telling stories that resonate with your audience's identity, you make them feel that your brand understands and represents who they are. This is why stories that focus on community, shared experiences, or common challenges can be so effective.

They make your customers feel that by choosing your brand, they are aligning with a community that reflects their own values and aspirations. Brené Brown, a researcher and storyteller, notes that "Stories are data with a soul," emphasizing how narratives can convey deeper meaning and foster a sense of belonging.

Simplifying Complex Information

Storytelling also helps to simplify complex information. When you need to convey a lot of data or explain a complicated concept, wrapping it in a story can make it more accessible and engaging.

For example, if you're offering a financial service that involves multiple steps or technical jargon, you can create a story about a character who navigates these complexities, making it relatable and easier to understand.

This approach not only keeps your audience engaged but also helps them remember the information better, as stories are easier to recall than dry facts or instructions.

A Nielsen study found that consumers are more likely to engage with brands that tell stories, with storytelling increasing recall by up to 22 times compared to facts alone. This demonstrates the power of storytelling in making complex information more digestible and memorable.

Enhancing the Customer Journey

A well-crafted story also enhances the customer's journey. From the first interaction to the final purchase, storytelling can guide your audience through each stage, making the process feel more natural and engaging. For instance, your website could tell the story of a customer's journey, from discovering a problem, finding your product as the solution, and then experiencing the benefits.

Each section of your site can represent a different chapter of this story, guiding the visitor toward the ultimate conclusion—making a purchase. This narrative structure makes the buying process feel less like a transaction and more like a satisfying conclusion to a story.

Differentiating Your Brand and Strengthening Calls to Action

In addition to building an emotional connection, storytelling is also a powerful way to differentiate your brand. In a crowded market, where many products offer similar features and benefits, the story you tell can set you apart.

A compelling narrative about your brand's origins, your commitment to quality, or the unique way your product is made can give customers a reason to choose you over competitors.

It's not just about what you sell, but about the story behind it, and how that story aligns with the values and aspirations of your audience. Storytelling can also enhance the effectiveness of your calls to action (CTAs).

When your CTA is framed within a story, it feels more like a natural next step in the narrative rather than a hard sell. For example, after telling the story of a customer who overcame a challenge using your product, your CTA could invite the reader to start their own journey by purchasing the product. This approach makes the CTA more compelling because it's tied to the emotional impact of the story, rather than being an isolated request.

Building Anticipation and Engagement

In content marketing, storytelling can also be used to build anticipation and engagement over time. By creating a series of stories or a narrative arc across multiple pieces of content, you can keep your audience coming back for more. Each piece of content adds a new chapter to the story, whether it's a blog post, an email, or a social media update, drawing your audience deeper into the narrative and strengthening their connection to your brand.

Conclusion

The power of storytelling in marketing lies in its ability to make your message more memorable, relatable, and impactful. As Maya Angelou famously said, *"People will forget what you said, people will forget what you did, but people will never forget how you made them feel."* A story doesn't just tell; it shows. It brings your brand to life and makes it resonate on a personal level with your audience.

When done well, storytelling transforms your marketing from a series of pitches into a meaningful experience that your customers will remember, relate to, and act upon. It's not just about crafting a narrative; it's about crafting a narrative that converts.

Call to Action (CTA)

Ready to harness the power of storytelling in your marketing?

Start by identifying the core stories that resonate with your audience—stories of transformation, success, or shared values.

Begin crafting these narratives today and watch as they create deeper connections with your customers, driving engagement and boosting conversions. Let your story be the one that makes a difference!

~CHAPTER 8~

Authority

Building Trust Through Expertise and Endorsements

Authority is a critical component of trust in marketing. When customers see you as an expert in your field, they're more likely to believe in the value of your products or services and take action based on your recommendations. Authority can be established in various ways, from showcasing your expertise to leveraging endorsements from respected figures in your industry. The goal is to create a perception that your brand is credible, reliable, and knowledgeable, making it easier for customers to trust you and make purchasing decisions.

Demonstrating Expertise Through Content

Building authority starts with demonstrating your expertise. This can be done through content that educates, informs, and provides value to your audience. Research shows that 96% of B2B buyers want content with more input from industry thought leaders. When you consistently produce high-quality content that addresses your audience's needs and challenges, you position yourself as a go-to resource in your niche.

For example, if you're in the weight loss industry, publishing in-depth guides on nutrition, exercise, and sustainable weight management strategies not only helps your audience

But also reinforces your authority as a knowledgeable source. Over time, this content builds a foundation of trust, making your audience more likely to turn to you when they're ready to make a purchase.

Leveraging Credentials and Experience

Authority is also established through the credentials and experience you bring to the table. Highlighting your qualifications, certifications, and years of experience can strengthen your credibility. A study by Nielsen found that 85% of consumers said they trust a product more when an expert recommends it, underscoring the value of perceived expertise.

If you have specialized knowledge or have been featured in reputable publications, these details should be prominently displayed. For instance, if you're offering financial advice, showcasing your certifications or your appearances in industry-leading media outlets can reassure potential customers that you're a trusted expert in the field. This credibility reduces the perceived risk of doing business with you and makes your audience more comfortable following your advice.

The Power of Endorsements

Endorsements from respected figures in your industry can also significantly boost your authority. When an influencer, expert, or well-known brand publicly supports your product, it serves as a powerful form of social proof. Their endorsement acts as a transfer of trust, as their followers or audience are likely to view you in the same positive light.

A study by the *Journal of Marketing Research* found that endorsements from credible influencers could increase purchase intent by up to 50%. If you're in the survival niche, for example, having a well-known survivalist endorse your gear can elevate your brand's credibility and attract customers who trust that expert's judgment. This type of authority by association can be incredibly effective in influencing purchasing decisions.

Communicating with Authority

Authority isn't just about what you say; it's also about how you say it. The tone and style of your communication can impact how your audience perceives your expertise. A confident, clear, and professional tone conveys that you know your subject matter well. Avoiding vague language or unnecessary jargon also helps to build trust, as it shows that you're capable of explaining complex topics in a way that's easy to understand. This clarity reassures your audience that they're in good hands and that they can rely on your advice and products.

Using Success Stories and Case Studies

Another way to establish authority is by sharing success stories and case studies. When you provide evidence of how your product or service has helped others achieve their goals, you're not just telling your audience that you're an expert—you're showing them. These real-world examples serve as proof that your expertise translates into tangible results.

For instance, if you're selling marketing services, a case study detailing how you helped a client significantly increase their online presence and sales can be a powerful testament to your authority. This evidence-based approach makes it easier for potential customers to trust you and feel confident in their decision to work with you.

Publishing on Reputable Platforms

Publishing in reputable platforms or contributing to well-known industry blogs can also enhance your authority. When your content is featured in respected publications, it signals to your audience that your insights are valued by other experts in the field. This external validation further solidifies your position as a thought leader and makes your brand more trustworthy.

If your expertise is recognized by others, it's easier for your audience to view you as an authority, which can lead to increased trust and higher conversion rates.

According to a study by Demand Gen Report, 47% of B2B buyers consume three to five pieces of content before engaging with a sales rep, often relying on content published on reputable platforms.

Building Authority Through Associations

In addition to showcasing your own authority, you can build trust by associating with other credible brands or individuals. Partnerships, collaborations, and joint ventures with respected names in your industry can enhance your reputation by association. When customers see you working alongside other trusted figures, they're more likely to view you as trustworthy as well. This is particularly effective in industries where trust is a critical factor in purchasing decisions, such as health, finance, or security.

Utilizing Reviews and Testimonials

Reviews and testimonials are another form of endorsement that can reinforce your authority. When satisfied customers publicly share their positive experiences with your brand, it not only builds social proof but also affirms your expertise. Potential customers are more likely to trust your brand when they see that others have had successful outcomes.

This is why it's important to actively collect and showcase testimonials from happy customers, especially those who are respected or influential in your niche. Their words can carry significant weight and help persuade others to follow their lead.

Overcoming Objections with Authority

Authority also plays a crucial role in overcoming objections. When potential customers have doubts or concerns, your established authority can help address and dispel those fears. For example, if a customer is hesitant about the effectiveness of a product, your demonstrated expertise and the endorsements you've received can reassure them that they're making the right choice.

This ability to alleviate concerns through authority can be the deciding factor in whether a customer moves forward with a purchase. According to a study by *MarketingProfs*, 70% of consumers feel more comfortable buying a product after they have read or viewed positive reviews and endorsements from experts.

Conclusion

Ultimately, authority in marketing is about building a foundation of trust that makes it easier for your audience to say yes to your offers. It's not just about claiming expertise; it's about proving it through consistent, high-quality content, credible endorsements, and a strong, confident presence. As Stephen Covey, author of *The 7 Habits of Highly Effective People*, said,

"Trust is the glue of life. It's the most essential ingredient in effective communication. It's the foundational principle that holds all relationships."

When your audience views you as an authority, they're more likely to listen to your recommendations, trust your products, and remain loyal to your brand.

By focusing on building and reinforcing your authority, you create a solid foundation for long-term success and deeper customer relationships.

Call to Action (CTA)

Ready to establish yourself as an authority in your field?

Start by creating valuable content, seeking endorsements

from respected figures and showcasing your credentials and success stories. Let your expertise shine and build the trust that converts!

Begin today and watch your authority grow alongside your business.

~CHAPTER 9~

Effective Digital Packaging

How Presentation Drives Perception and Sales

Effective digital packaging is more than just wrapping your product in a visually appealing design. It's about shaping how your audience perceives your brand and influencing their decision to buy. In the digital world, where customers can't physically touch or feel a product, the way you present it online becomes a critical factor in driving sales. Digital packaging combines visual design, messaging, and overall user experience to create an impression that resonates with your audience and convinces them of your product's value.

The Power of First Impressions

At the core of effective digital packaging is the concept of first impressions. Research indicates that it takes about 50 milliseconds (0.05 seconds) for users to form an opinion about your website that determines whether they'll stay or leave.

This initial perception often dictates whether they'll stay to learn more or click away to explore other options. By ensuring that your digital packaging is not only visually striking but also aligned with your brand message, you can capture attention and spark interest immediately.

For example, if you're selling a premium weight loss supplement, your product page should exude luxury and effectiveness, from the sleek design to the compelling copy that highlights its benefits.

The Role of Visual Design

Visual design plays a crucial role in digital packaging. It's not just about looking good; it's about communicating your brand's identity and the value of your product at a glance. A study by the Stanford Persuasive Technology Lab found that 46.1% of participants assessed the credibility of websites based on their visual design. Colors, typography, and imagery all work together to tell a story about your product.

If you're in the survival niche, using earthy tones, rugged fonts, and images of nature can create a sense of adventure and reliability, making your product feel essential for anyone who values preparedness. The goal is to ensure that the visual elements of your packaging evoke the right emotions and align with the expectations of your target audience.

High-Quality Product Imagery

The product image is another key component of digital packaging. High-quality, detailed images can make your product more tangible in a digital space.

Customers want to see what they're getting, and the more they can visualize themselves using your product, the more likely they are to buy. Research shows that online shoppers think that the quality of a product's image is more important than product-specific information, long descriptions, or even reviews.

Offering multiple angles, zoom features, and even videos can help bridge the gap between the digital and physical experience. If you're selling a piece of survival gear, showing it in action—perhaps being used in a wilderness setting—can make it more appealing and give customers confidence in its practicality and durability.

Crafting Compelling Messaging

Messaging is equally important in digital packaging. The way you describe your product, its features, and its benefits can either reinforce its perceived value or undermine it. Effective messaging focuses on what matters most to your customers, addressing their pain points and highlighting how your product solves them.

For instance, if you're marketing a new financial planning tool, your messaging should emphasize how it simplifies the complex task of managing finances, saves time, and helps users achieve their financial goals. The key is to be clear, concise, and customer-focused, ensuring that your message resonates with what your audience cares about most. As noted by legendary advertising executive David Ogilvy, *"If it doesn't sell, it isn't creative."* The creativity in your messaging must always serve the purpose of driving conversions.

Enhancing User Experience

Digital packaging also involves the overall user experience on your website or online store. A cluttered, confusing layout can detract from even the most beautifully designed product page. On the other hand, a clean, intuitive design enhances the digital packaging by making it easy for customers to find the information they need, navigate between pages, and make a purchase. Streamlining the checkout process, offering clear calls to action, and ensuring fast load times all contribute to a positive user experience that supports the effectiveness of your digital packaging.

Leveraging Social Proof

Another aspect of digital packaging is the use of social proof. Incorporating customer reviews, ratings, and testimonials directly into your product pages can boost credibility and make your product more appealing. Research by Spiegel Research Center shows that displaying reviews can increase conversion rates by up to 270%, especially when customers are unfamiliar with the product. When potential customers see that others have had positive experiences with your product, it reinforces their perception of its value and can be the nudge they need to complete a purchase.

Strategic Pricing Presentation

Pricing presentation is also a critical element of digital packaging. How you display the price can impact the customer's perception of value. If your product is premium, the price should be positioned in a way that reinforces its quality and exclusivity. This might involve using a sleek, minimalistic design around the price or highlighting any discounts or value propositions that justify the cost. On the other hand, for budget-friendly products, the focus might be on transparency and affordability, making sure the price is easy to find and understand. In either case, the way you package the price can either enhance or diminish the perceived value of your product.

The Benefits of Bundling

Bundling products together is another effective digital packaging strategy. Offering a package deal can make your customers feel like they're getting more value for their money. Research from the Harvard Business Review suggests that bundling can increase sales by creating a perceived sense of value and exclusivity. For instance, if you're in the health and wellness niche, you might bundle a main product like a dietary supplement with a complementary item, such as a meal plan or workout guide.

This not only increases the perceived value but also encourages higher spending per transaction. Presenting these bundles attractively, with clear descriptions of each component and the benefits of purchasing them together, can drive sales and enhance the customer's overall experience.

Consistency Across Platforms

Digital packaging also extends to how your brand is perceived across different platforms. Consistency in design and messaging across your website, social media, and email marketing reinforces your brand identity and helps build trust with your audience. When customers encounter your brand in different places online, they should experience the same high-quality presentation that reinforces their positive perception of your products. This consistent packaging across platforms ensures that your brand remains top of mind and continues to resonate with your audience wherever they engage with you.

Conclusion

Effective digital packaging is about more than just aesthetics. It's a strategic approach to presenting your product in a way that maximizes its perceived value and drives customer action.

By focusing on visual design, high-quality imagery, compelling messaging, and a seamless user experience, you create a digital package that not only attracts attention but also converts it into sales. As branding expert Marty Neumeier puts it, "A brand is not what you say it is. It's what they say it is." Your digital packaging plays a crucial role in shaping that perception.

Call to Action (CTA)

Ready to elevate your digital packaging and boost your sales?

Start by analyzing your current product presentation, from visuals to messaging, and see where you can make improvements.

Focus on creating a cohesive, compelling package that resonates with your audience and drives them to take action. Make your digital presence as irresistible as your product—begin your packaging makeover today!

~CHAPTER 10~

Framing

Shaping Customer Choices Through Strategic Context

Framing is a powerful tool in marketing that shapes how your customers perceive your products or services. It's not about changing the facts but about presenting them in a way that influences how those facts are interpreted. The way you frame information can significantly impact your customers' decisions, guiding them toward the choices you want them to make. By strategically presenting your offers, you can create a context that makes your products more appealing and drives higher conversion rates.

The Psychology Behind Framing

At its core, framing is about context. The same product can be perceived differently depending on how it's presented. Research by Nobel Prize-winning psychologists Daniel Kahneman and Amos Tversky found that people's decisions are heavily influenced by how choices are framed.

For example, they demonstrated that people are more likely to opt for a treatment when told it has a "90% survival rate" than when told it has a "10% mortality rate," even though both statements are statistically equivalent. This highlights the importance of framing in shaping perceptions and influencing decisions.

For instance, if you're selling a weight loss program, framing it as "Lose 10 pounds in 30 days" versus "Achieve a healthier you in 30 days" can lead to different customer reactions.

The first frame emphasizes a specific, tangible result, appealing to those who are focused on quick outcomes.

The second frame highlights overall well-being, which might attract customers who are more concerned with long-term health. Both statements are true, but the frame you choose will influence who finds the offer more compelling.

Strategic Pricing Frames

One of the most common ways to use framing in marketing is through pricing strategies. The way you present the price can make a big difference in how customers perceive the value of your product. According to a study by *The Journal of Consumer Research*, consumers are more likely to perceive a product as a good deal when it is framed positively, such as "Save $20" versus "Get 20% off," even if both discounts result in the same final price .

The word "only" or the phrase "under $100" frames the price in a way that emphasizes affordability, making the price point feel more accessible. This subtle shift in language can significantly impact the customer's perception of the offer's value.

Framing Benefits to Match Customer Desires

Framing also plays a crucial role in how you present product benefits. Instead of simply listing features, you can frame them in a way that aligns with your customers' desires and pain points. For instance, in the survival niche, rather than just stating that a piece of gear is "waterproof," you might frame it as "Keeps your equipment safe and dry in the harshest conditions."

This framing shifts the focus from a technical specification to a real-world benefit that resonates more deeply with the customer's needs. It's not just about what the product does; it's about how it will improve the customer's life.

The Decoy Effect in Choice Architecture

Another effective use of framing is in the presentation of choices. When offering multiple options, how you position each choice can guide customers toward the one you want them to choose. This is often referred to as the "decoy effect." Research from the *Journal of Marketing* shows that introducing a third, less attractive option (a decoy) can nudge customers toward a more profitable option.

For example, if you're offering a subscription service, you might present three pricing tiers: a basic, a standard, and a premium. By framing the standard option as the best value—perhaps by highlighting it in the middle and adding a "most popular" label—you make it more attractive compared to the basic option, while the premium option makes the standard seem more reasonably priced. This subtle framing encourages customers to gravitate toward the option that provides the best perceived value, even if it's not the cheapest.

Overcoming Objections with Framing

Framing is also effective in addressing customer objections. When potential customers hesitate due to concerns about cost, commitment, or effectiveness, the way you frame your response can either alleviate or exacerbate those concerns. For example, if you're selling a high-ticket item, framing the investment as a long-term benefit—such as "An investment in your health that will pay off for years to come"—can shift the focus from the immediate cost to the enduring value.

This helps customers see the purchase as a smart decision rather than a financial burden, increasing the likelihood of conversion.

Enhancing Social Proof Through Framing

The power of framing extends to social proof as well. How you present testimonials, reviews, and endorsements can influence how customers perceive the credibility of your product. For instance, framing a testimonial by highlighting a specific result— *"John lost 15 pounds in 6 weeks using our program"*—focuses on a concrete outcome, making it more persuasive. Alternatively, framing a testimonial around the overall experience—*"This program made losing weight enjoyable and sustainable"*—can appeal to customers who are more interested in the process than just the results.

A study by the *International Journal of Hospitality Management* found that framing customer reviews in a positive light significantly enhanced perceived service quality and customer satisfaction.

Framing Your Brand Narrative

Framing also plays a role in the overall narrative you create around your brand. How you frame your brand's story, mission, and values can attract different types of customers and foster deeper connections. If you frame your brand as a pioneer in innovation, you attract early adopters and tech enthusiasts.

If you frame it as a champion of sustainability, you appeal to environmentally conscious consumers. The key is to ensure that the framing aligns with your target audience's values and aspirations, making your brand more relatable and trustworthy.

Content and Email Marketing Frames

In digital marketing, framing can be applied to the way you structure your content. Headlines, for example, are a crucial element of framing. A headline like "10 Simple Steps to Financial Freedom" frames the content as easy and achievable, making it more attractive to readers looking for practical advice. On the other hand, "10 Expert Tips for Building Wealth" frames the content as authoritative and valuable, appealing to those seeking expert guidance.

Both headlines promise similar content, but the framing targets different motivations, influencing which audience will be more likely to engage. In email marketing, the subject line you choose can frame the content of the email in a way that increases open rates. For example, framing an email subject as *"Unlock Exclusive Savings Today"* creates a sense of urgency and exclusivity, making it more likely to be opened.

Conclusion

Framing is a strategic tool that can profoundly shape customer perceptions and choices. By carefully considering how you present your product, pricing, benefits, and brand, you can influence the way customers view your offerings and guide them toward favorable decisions. Effective framing helps you create a narrative that resonates with your audience, aligns with their values, and makes your product the clear choice. As renowned psychologist and author Daniel Kahneman said, *"Nothing in life is as important as you think it is, while you are thinking about it."* Framing ensures that what your customers think about your product is positive and compelling.

Call to Action (CTA)

Ready to leverage the power of framing in your marketing strategy?

Start by analyzing your current messaging and presentation to identify opportunities for reframing. Experiment with different frames to see what resonates most with your audience and watch as your conversions and customer satisfaction grow.

Begin shaping perceptions today with the strategic use of framing!

~CHAPTER 11~

Consistency and Commitment:

Turning Small Yeses into Big Wins

Consistency and commitment are two powerful principles in marketing that can turn small initial actions into significant customer loyalty and sales. These concepts are grounded in the idea that once people commit to something, even in a small way, they are more likely to continue in that direction to remain consistent with their initial decision. This psychological tendency can be leveraged in your marketing strategy to build deeper customer relationships and drive long-term success.

The Psychology of Consistency and Commitment

At the heart of the consistency principle is the understanding that people like to be seen as consistent in their behavior, values, and decisions. Psychologist Robert Cialdini, in his book *Influence: The Psychology of Persuasion*, explains that people have a deep-seated desire to be consistent with their commitments. This drive for consistency stems from the social pressure to appear reliable and trustworthy, both to oneself and to others.

Once a customer makes a small commitment—whether it's signing up for a newsletter, following you on social media, or making a low-cost purchase—they're more likely to take further actions that align with that initial choice.

This is because consistency is tied to self-image; once someone sees themselves as the kind of person who supports your brand or values your product, they want to maintain that identity.

The Power of Small Commitments

The power of commitment can be seen in the classic marketing tactic of the "foot-in-the-door" technique. This strategy involves getting customers to agree to a small request first, which then increases the likelihood that they'll agree to a larger request later. Research published in the *Journal of Personality and Social Psychology* shows that even small initial commitments can lead to significant changes in behavior.

For example, if you're marketing a survival kit, you might start by offering a free guide on emergency preparedness. Once a customer has downloaded the guide, they've made a small commitment to your brand. When you later offer a comprehensive survival kit, the customer is more likely to purchase it because they've already engaged with your content and see themselves as someone who values preparedness.

Subscription Models and Free Trials

This approach is particularly effective in subscription-based businesses. Offering a free trial or a low-cost entry point allows customers to make an initial commitment with minimal risk.

Research by *Invesp* shows that approximately 60% of customers will decide to continue using a subscription service after a free trial period, indicating that the initial small commitment can lead to a more significant long-term decision. Once they've started using the service and experienced its benefits, they're more likely to upgrade to a paid plan or renew their subscription.

The initial "yes" leads to a pattern of behavior that aligns with the commitment they've already made, turning a small action into a long-term customer relationship.

Building Loyalty Through Consistent Engagement

Consistency and commitment are also crucial in content marketing. When you consistently deliver valuable content, you build a relationship with your audience that encourages ongoing engagement. If you're in the weight loss industry, regularly sharing tips, recipes, and success stories keeps your audience coming back, reinforcing their commitment to your brand. Over time, this consistent engagement can lead to higher conversion rates, as your audience becomes more invested in what you offer. The small, regular interactions add up, creating a strong foundation of trust and loyalty.

Personalized Follow-Ups and Customer Journeys

Another way to leverage consistency is through personalized follow-ups and customer journeys. After a customer makes a purchase or signs up for a service, sending them personalized recommendations or related offers can encourage them to continue engaging with your brand.

For example, if a customer buys a beginner's fitness program, following up with an offer for a more advanced program or complementary products keeps them on the path they've started. This approach not only increases sales but also reinforces the customer's commitment to their fitness goals, which they associate with your brand.

Utilizing Social Proof and Testimonials

Consistency and commitment also play a role in social proof and testimonials. When potential customers see that others have consistently chosen your product and committed to it over time, they're more likely to do the same.

A study by *BrightLocal* found that 88% of consumers trust online reviews as much as personal recommendations, highlighting the importance of consistent positive feedback from satisfied customers. Showcasing long-term customer stories, where users talk about how they've stuck with your product and the benefits they've gained, can be a powerful motivator. It taps into the desire to be consistent with what others in the community are doing, encouraging new customers to follow suit.

Effective Email Marketing Strategies

In email marketing, the principles of consistency and commitment can be used to build deeper engagement over time. A welcome series of emails, for instance, gradually introduces new subscribers to your brand, starting with a simple thank you for signing up, followed by useful content and small offers. Each email builds on the last, reinforcing the subscriber's decision to join your list and encouraging further actions, like making a purchase or sharing your content. This steady progression from small commitments to bigger ones helps turn subscribers into loyal customers.

The Role of Loyalty Programs

These principles are also effective in loyalty programs. When customers commit to a loyalty program, they're more likely to continue purchasing from you to maintain their status or earn rewards. According to a study by *Bond Brand Loyalty*, 71% of consumers say loyalty programs are a meaningful part of their brand relationships.

The initial commitment of joining the program, combined with the ongoing consistency of earning points or benefits, creates a cycle of repeated engagement. This not only increases customer retention but also enhances their overall lifetime value. The small, consistent actions—like making regular purchases to earn rewards—lead to significant long-term wins for your business.

Gradual Escalation in Calls to Action

Another application of consistency and commitment is in the way you frame your calls to action. Instead of immediately pushing for a large commitment, like a high-ticket purchase, start with something smaller and easier to agree to. Once the customer has taken that first step, they're more likely to take the next one, and the one after that. This gradual escalation, from small yeses to bigger commitments, aligns with their desire to be consistent in their choices and actions.

Conclusion

The effectiveness of consistency and commitment is rooted in human psychology. People want to align their actions with their past behaviors and the commitments they've made. When you understand this, you can create marketing strategies that guide customers along a path from initial interest to long-term loyalty. As philosopher William James said, "Act as if what you do makes a difference. It does." Every small commitment made by a customer can lead to a larger impact for your business.

Call to Action (CTA)

Ready to harness the power of consistency and commitment in your marketing strategy? Start by identifying small, low-risk commitments you can ask your customers to make today. Then, build on those initial actions with consistent engagement and personalized follow-ups.

Watch as these small yeses turn into big wins, driving customer loyalty and boosting your sales. Begin today by creating a path of commitments that lead to long-term succeed.

~CHAPTER 12~

Customer Loyalty

Strategies for Cultivating Lifelong Fans

Customer loyalty is the foundation of a successful business. It's not just about making a one-time sale; it's about building a relationship with your customers that keeps them coming back, time and time again. Loyal customers are more valuable than new ones—they're more likely to make repeat purchases, refer others to your business, and advocate for your brand.

In fact, research from Bain & Company shows that increasing customer retention rates by 5% can increase profits by 25% to 95%. Cultivating customer loyalty involves more than just offering a good product or service; it requires a strategic approach to create lasting connections that turn one-time buyers into lifelong fans.

Exceptional Customer Service

One of the most effective ways to build customer loyalty is through exceptional customer service. When customers feel valued and supported, they're more likely to remain loyal to your brand.

According to a study by HubSpot, 93% of customers are likely to make repeat purchases with companies that offer excellent customer service. This means going beyond just resolving issues; it's about creating positive experiences at every touch point.

Whether it's a prompt response to a query, a personalized thank you note, or a follow-up to ensure satisfaction, each interaction should reinforce the customer's needs, and satisfaction are your top priority. Over time, these small but meaningful actions add up, fostering a sense of trust and loyalty.

The Power of Personalization

Another key strategy for cultivating loyalty is personalization. Customers appreciate it when a brand understands their preferences, needs, and behaviors. By leveraging data and insights, you can tailor your offerings, communications, and recommendations to each customer's unique profile. A study by Epsilon found that 80% of consumers are more likely to purchase from a brand that provides personalized experiences.

For example, if you're in the weight loss industry, sending personalized meal plans, workout routines, or product recommendations based on a customer's goals can make them feel understood and valued. This personalized approach not only enhances the customer's experience but also strengthens their connection to your brand, making them more likely to stay loyal.

Effective Loyalty Programs

Loyalty programs are a tried-and-true method of encouraging repeat business. These programs reward customers for their ongoing patronage, whether through points, discounts, exclusive offers, or other incentives. The key to a successful loyalty program is to ensure that it offers real value to the customer. Research by Bond Brand Loyalty indicates that 73% of consumers are more likely to recommend brands with good loyalty programs.

For instance, a survival gear store might offer a points-based system where customers earn points for every purchase, which they can then redeem for discounts on future purchases or exclusive gear. The more value customers perceive in the program, the more likely they are to participate actively and continue choosing your brand over competitors.

Building a Community

Building a community around your brand is another powerful way to foster customer loyalty. When customers feel like they're part of a group that shares their values and interests, they're more likely to stay engaged with your brand. According to a report by Sprout Social, 57% of consumers are more likely to increase their spending with a brand when they feel part of a community. This can be achieved through social media groups, forums, or even regular events or webinars where customers can connect with each other and with your brand.

For example, a brand in the survival niche might host monthly webinars on outdoor skills or emergency preparedness, bringing together customers who share a passion for these topics. This sense of community not only enhances loyalty but also turns your customers into advocates who promote your brand to others.

Consistency in Branding and Messaging

Consistency in your branding and messaging is also crucial for building loyalty. Customers need to know what to expect from your brand, whether it's the quality of your products, the tone of your communications, or the values you stand for. When your brand consistently delivers on its promises, it builds trust, and trust is the bedrock of loyalty. A study by Lucidpress found that consistent branding across all channels can increase revenue by up to 23%.

For example, if your brand is known for high-quality, reliable products, consistently delivering on that promise reinforce the customer's decision to keep coming back. This consistency helps cement your brand in the customer's mind as the go-to choice for their needs.

Offering Exclusive Content and Offers

Exclusive content and offers can also play a significant role in fostering loyalty. When customers feel like they're getting something special that isn't available to everyone, it creates a sense of privilege and appreciation. This can be as simple as offering early access to new products, special discounts, or exclusive content that provides added value. In the weight loss industry, for example, offering members-only recipes, tips, or access to exclusive webinars can make customers feel valued and encourage them to remain loyal to your brand.

Transparency and Honesty

Transparency and honesty are vital in maintaining customer loyalty. Customers value brands that are open about their practices, policies, and product offerings. Being transparent about where your products come from, how they're made, or how you handle customer data can build trust and loyalty. According to a survey by Label Insight, 94% of consumers are likely to be loyal to a brand that offers complete transparency.

When customers know they can rely on you to be honest and straightforward, even when things go wrong, they're more likely to stay loyal in the long term.

Providing Continuous Value

Another way to cultivate loyalty is by consistently delivering value beyond the product itself. This can include providing helpful content, resources, or education that enhances the customer's experience with your product. For instance, if you're in the survival niche, offering a series of educational emails on survival skills or how to use the gear effectively can add significant value to the customer's purchase. This approach not only keeps the customer engaged but also reinforces their decision to buy from you, increasing the likelihood that they'll return for future purchases.

Listening to Customer Feedback

Listening to customer feedback and acting on it is another crucial component of building loyalty. When customers see that you're paying attention to their needs and making improvements based on their input, it strengthens their bond with your brand. According to Gartner, when brands improve the quality of their customer feedback programs, they achieve 10% greater customer retention rates.

This could involve making changes to your products based on customer suggestions, improving your website's user experience, or even just acknowledging and thanking customers for their feedback. This level of responsiveness shows that you value your customers' opinions, which can go a long way in fostering loyalty.

Aligning with Shared Values

Creating a sense of shared values between your brand and your customers can deepen loyalty. When customers feel that your brand stands for something they believe in, they're more likely to stick with you.

Whether it's a commitment to sustainability, social responsibility, or quality, aligning your brand with values that resonate with your customers can create a powerful, lasting connection. A study by Edelman found that 64% of consumers worldwide make purchasing decisions based on a brand's social or political stance. In the weight loss industry, for example, promoting a message of healthy, sustainable weight management rather than quick fixes can attract customers who share those values and are more likely to remain loyal.

Conclusion

Customer loyalty isn't something that happens overnight. It's built through consistent, thoughtful actions that show your customers they're valued and understood. As Maya Angelou wisely said, "People will forget what you said, people will forget what you did, but people will never forget how you made them feel." By creating positive experiences and meaningful connections, you can turn one-time buyers into lifelong fans who not only keep coming back but also spread the word about your brand.

Call to Action (CTA)

Ready to start building a loyal customer base that drives long-term success? Begin by focusing on personalized service, consistent communication, and creating a sense of community around your brand. Show your customers that they are valued and that you stand for something they believe in. Start today and watch your loyal customer base grow!

~CHAPTER 13~

Decision Fatigue

Simplifying Choices to Boost Conversions

Decision fatigue is a phenomenon that occurs when people become overwhelmed by the sheer number of choices they have to make, leading to reduced decision-making quality and often, decision paralysis. In the context of online marketing, decision fatigue can be a significant barrier to conversions. When customers are presented with too many options, they may struggle to make a choice, or worse, abandon the decision altogether. Simplifying choices for your customers is key to overcoming decision fatigue and boosting conversions.

Understanding Decision Fatigue

The concept of decision fatigue was popularized by social psychologist Roy F. Baumeister, who found that making decisions depletes a person's mental energy, leading to poorer decision quality over time. This mental depletion can cause people to avoid making decisions altogether, opt for the easiest choice, or revert to a default option.

A study published in the *Proceedings of the National Academy of Sciences* highlighted that judges were more likely to grant parole at the beginning of the day or right after a meal break, illustrating how decision fatigue can lead to less favorable outcomes due to depleted mental energy. or overly complex options can actually deter customers from making a purchase Reducing decision fatigue involves simplifying the decision-making process, so customers feel more confident and are less likely to experience overwhelm.

Reducing the Number of Choices

The first step in addressing decision fatigue is to reduce the number of options you present to your customers. Research from the *Journal of Personality and Social Psychology* found that when people are presented with fewer choices, they are more likely to make a decision and be satisfied with their choice. While it might seem logical to offer a wide range of choices to cater to different preferences, too many options can lead to confusion and hesitation. Instead, focus on curating a selection of your best products or services.

For example, if you're running an online store that sells survival gear, rather than offering 20 different types of water filters, you could highlight the top three that are most popular or highly rated. This not only simplifies the decision-making process but also helps guide customers toward the options that are most likely to meet their needs.

Clear Categorization and Navigation

Another effective strategy is to use clear and concise categorization. When customers can easily navigate your offerings based on their specific needs, it reduces the mental effort required to make a decision. For instance, if you're selling weight loss supplements, organizing them into categories like "Appetite Suppressants," "Metabolism Boosters," and "Energy Enhancers" can help customers quickly identify the type of product that aligns with their goals.

This approach makes the selection process feel more manageable and less overwhelming, encouraging customers to move forward with their purchase. According to a study by the *Baymard Institute*, 61% of customers cite poor navigation and search functions as a reason for abandoning a website, emphasizing the importance of clear categorization.

Highlighting Recommendations and Social Proof

Presenting a clear, standout recommendation is another way to combat decision fatigue. By highlighting a "best seller," "most popular," or "expert's choice," you provide a shortcut for customers who might be uncertain about which option to choose. This not only helps reduce the cognitive load but also instills confidence in their decision. If a customer sees that a particular survival kit is labeled as the "best value," they're more likely to trust that recommendation and make a purchase without second-guessing themselves. This approach leverages social proof and authority to guide the decision-making process.

Simplifying Product Information

In addition to reducing the number of choices, simplifying the information presented with each option is crucial. Overloading customers with details, specifications, or features can contribute to decision fatigue. Instead, focus on the most important benefits and key differentiators that matter to your audience.

For example, if you're selling a fitness tracker, rather than listing every single feature, you might highlight three key benefits like heart rate monitoring, sleep tracking, and water resistance. This concise presentation helps customers quickly understand the value of the product without feeling overwhelmed by unnecessary information.

Streamlining the Checkout Process

Streamlining the checkout process is another essential aspect of minimizing decision fatigue. The more steps or decisions a customer has to make during checkout, the more likely they are to experience fatigue and abandon the process.

According to the *Baymard Institute*, the average cart abandonment rate is around 69%, often due to a complicated checkout process. Simplifying the checkout by reducing the number of fields, offering guest checkout options, and providing clear, step-by-step instructions can make the process smoother and more straightforward. When customers encounter fewer obstacles, they're more likely to complete their purchase without hesitation.

Bundling and Simplified Offers

Bundling products together is another strategy that can help reduce decision fatigue. By offering pre-selected bundles or packages, you take the guesswork out of the equation for your customers. A study by *Harvard Business Review* found that bundling can increase perceived value and simplify decision-making, leading to higher sales. For instance, if you're selling camping gear, offering a "Complete Camping Kit" that includes a tent, sleeping bag, and cooking equipment can simplify the purchasing decision. Customers don't have to worry about selecting each item individually; they can trust that the bundle has everything they need, making the buying process more convenient and appealing.

Visual Hierarchy and Design

Visual hierarchy also plays a role in reducing decision fatigue. The way you design your website and present options can guide customers' eyes and help them focus on the most important elements. Using larger fonts, bold colors, or prominent placement for key choices can draw attention and make the decision-making process more intuitive.

For example, if you want to promote a specific product or offer, placing it at the top of the page with a clear call to action can help steer customers toward that option without overwhelming them with too many competing elements.

Offering Risk Mitigation Options

Offering a trial period or money-back guarantee can also alleviate decision fatigue by reducing the perceived risk of making a wrong choice. A study from the *Journal of Retailing* suggests that risk-reducing policies, such as money-back guarantees, can increase consumer confidence and willingness to purchase. When customers know they can try a product and return it if it doesn't meet their expectations, they're more likely to make a decision without the fear of regret. This approach can be particularly effective for high-cost or complex products where customers might otherwise hesitate to commit.

Creating a Sense of Urgency

Another effective tactic is to limit time or availability, creating a sense of urgency that pushes customers to make a decision more quickly. However, this should be done carefully to avoid adding pressure that could exacerbate decision fatigue. The goal is to motivate action without overwhelming the customer. For instance, a limited-time discount or a countdown timer for a special offer can encourage customers to make a decision sooner rather than later, helping to overcome hesitation and drive conversions.

Providing Decision-Making Tools

Finally, providing helpful tools like comparison charts, quizzes, or guides can empower customers to make informed decisions without feeling overwhelmed. These tools simplify the process by breaking down options based on the customer's specific needs or preferences. According to a survey by *Salesforce*, 62% of consumers say that a brand's online content plays a key role in influencing their final purchasing decision.

For example, a quiz that helps customers choose the right weight loss supplement based on their lifestyle and goals can reduce the number of decisions they need to make, guiding them toward the best option without the stress of sorting through all the available products.

Conclusion

Minimizing decision fatigue is about making the buying process as simple and stress-free as possible. As Steve Jobs famously said, "Simple can be harder than complex: You have to work hard to get your thinking clean to make it simple." By reducing the number of choices, simplifying information, and streamlining the purchasing process, you can boost conversions and create a more positive experience for your audience. When customers feel confident and clear-headed in their choices, they're more likely to follow through with a purchase and return for future transactions.

Call to Action (CTA)

Ready to simplify your customers' decision-making process and boost conversions? Start by evaluating your current offerings and identify areas where you can reduce choice complexity, streamline navigation, and simplify checkout.

Implement these strategies today and watch as your conversion rates increase and your customers enjoy a more seamless shopping experience!

~CHAPTER 14~

Design

Crafting Visually Compelling Digital Experiences

Design plays a pivotal role in how your audience interacts with your brand. It's not just about aesthetics; it's about creating a visually compelling digital experience that draws people in, keeps them engaged, and ultimately drives them to take action. In an online environment where customers can't physically interact with products, design becomes the primary means of communication, conveying your brand's values, quality, and credibility at a glance. Effective design doesn't just look good—it serves a purpose, guiding users seamlessly through their journey from discovery to conversion.

Understanding the Impact of Design

A visually compelling design starts with understanding your audience. The colors, fonts, and images you choose should resonate with the people you're trying to reach. Research from the *Journal of Business Research* found that color increases brand recognition by up to 80% and can influence consumer buying decisions by up to 85%.

For instance, if you're in the weight loss industry, a clean and energetic design with bright, uplifting colors can evoke feelings of health and vitality, encouraging users to explore your-offerings On the other hand, if you're selling high-end survival gear, a more rugged design with earthy tones and strong typography can convey durability and reliability.

Every design choice should align with the expectations and desires of your target audience, making them feel at home as soon as they land on your site.

Consistency: Building Trust and Recognition

Consistency is key in design. When your design elements are consistent across all platforms—your website, social media, emails, and even packaging—it creates a cohesive brand experience that reinforces trust and recognition. According to a study by Lucidpress, consistent presentation of a brand has been seen to increase revenue by 33%. Customers should feel a sense of familiarity no matter where they encounter your brand. This consistency helps to build a strong brand identity that sticks in the minds of your audience, making it easier for them to recall and choose your brand in the future. It's about creating a unified visual language that speaks to your brand's essence and reinforces your message at every touch point.

Effective Navigation: Guiding the User Journey

Navigation is another crucial aspect of design. A visually compelling website that's difficult to navigate will frustrate users and drive them away. Simplicity and clarity are essential. Menus should be intuitive, and pathways to key actions—like making a purchase or signing up for a newsletter—should be clear and direct. Studies show that 94% of users say that easy navigation is the most important feature of a website.

For example, a well-placed call-to-action (CTA) button with a contrasting color can guide users exactly where you want them to go, whether that's adding a product to their cart or downloading a guide. Effective navigation design reduces friction, making it easy for users to move through your site and complete their desired actions.

Leveraging White Space for Clarity

The use of white space is another powerful tool in crafting a visually compelling design. White space, or negative space, isn't just empty space; it's a critical design element that helps to highlight important content and reduce clutter. By giving your content room to breathe, white space enhances readability and draws attention to key messages.

Research by Crazy Egg indicates that increasing white space around text and titles increases user attention by 20%. For example, if you're promoting a new product, surrounding the product image and description with ample white space can make it stand out more, making the design feel more elegant and focused. White space also contributes to a sense of sophistication and professionalism, which can elevate the perceived value of your brand.

Typography: Conveying Tone and Clarity

Typography also plays a significant role in design. The fonts you choose should not only be readable but should also convey the tone of your brand. Bold, modern fonts can communicate strength and innovation, while more classic, serif fonts might suggest tradition and reliability. Typography contributes significantly to first impressions, with the choice of font potentially increasing readability and user retention by up to 18%.

In the survival niche, for example, using sturdy, no-nonsense fonts can reinforce the rugged, dependable nature of your products. Consistent use of typography across your digital platforms helps to maintain brand identity and ensures that your content is easy to read and engaging.

The Power of Visuals

Images are another vital component of design. High-quality images that resonate with your audience can make a significant impact. According to a study by Venngage, visual content is more than 40 times more likely to get shared on social media than other types of content . Whether it's product photos, lifestyle images, or even custom illustrations, visuals should support your brand's message and appeal to your target audience's emotions.

If you're in the weight loss industry, showing images of healthy, vibrant individuals using your products can inspire potential customers and create a strong emotional connection. In contrast, in the survival niche, images of rugged landscapes or people using survival gear in extreme conditions can evoke a sense of adventure and preparedness. The right images not only grab attention but also tell a story that complements your content.

Strategic Layout for Engagement

The layout is equally important in crafting a visually compelling design. A well-structured layout guides the user's eye through your content in a logical and engaging way. Research suggests that people read websites in an "F" pattern, focusing first on the upper left corner, and then moving horizontally across the page. Placing the most important information where the eye naturally falls—typically the top left of the page or prominently in the center—can help capture and retain attention. A grid-based layout can help maintain order and consistency, ensuring that your design feels balanced and professional.

Whether you're designing a landing page, a product page, or an email, the layout should make it easy for users to digest information quickly and take the next step.

Enhancing Interactivity

Interactive elements can also enhance the design by making the user experience more engaging. Features like hover effects, clickable images, and animated transitions can add a layer of interactivity that keeps users interested and encourages them to explore further. According to a study by *Adobe*, websites with interactive content were perceived as more engaging by 70% of consumers.

However, it's important to use these elements sparingly and purposefully. Overloading a site with too many animations or interactive features can overwhelm users and detract from the main message. The goal is to enhance the user experience, not to distract from it.

The Importance of Responsive Design

The mobile experience is an essential consideration in design. With more people accessing the internet via smartphones and tablets, your design must be responsive meaning it adapts seamlessly to different screen sizes and devices. Google's research indicates that 61% of users are unlikely to return to a mobile site they had trouble accessing, and 40% visit a competitor's site instead.

A design that looks great on a desktop but is difficult to navigate on a mobile device will lose you potential customers. Responsive design ensures that your website is accessible and easy to use, no matter how your audience chooses to interact with it. This flexibility is crucial in today's digital landscape, where user experience across all devices can make or break a conversion.

Conclusion

Design is about more than just aesthetics; it's about creating an experience that resonates with your audience, communicates your brand's value, and drives action. As Steve Jobs famously stated, "Design is not just what it looks like and feels like. Design is how it works." By focusing on visual appeal, usability, and consistency, you can craft a visually compelling digital experience that not only attracts attention but also keeps customers engaged and encourages them to take the next step with your brand. A well-designed digital presence is a powerful tool in building a lasting connection with your audience and achieving your business goals.

Call to Action

Ready to elevate your digital design and boost engagement?

Start by reviewing your current design elements to ensure they align with your brand identity and resonate with your target audience. Focus on creating a clean, intuitive, and visually appealing experience that guides your customers seamlessly from discovery to conversion.

Begin your design transformation today and watch your business thrive!

~CHAPTER 16~

Gestalt Principles

Designing for How People Perceive

Gestalt principles are foundational concepts in design that focus on how people naturally perceive and process visual information. These principles are based on the idea that the human brain organizes visual elements into groups or unified wholes rather than seeing them as isolated parts. As the famous Gestalt psychologist Kurt Koffka once said, "The whole is other than the sum of its parts." By understanding and applying Gestalt principles, you can create designs that are intuitive, engaging, and effective in conveying your message. These principles help guide your audience's perception, making it easier for them to navigate your content and take the desired action.

Proximity: Creating Meaning Through Closeness

One of the key Gestalt principles is **proximity**. This principle suggests that elements that are close to each other are perceived as related or part of the same group. In design, you can use proximity to create associations between different elements.

For example, grouping related products together on an e-commerce site or placing a call-to-action button near a product description signals to the user that these elements are-connected This principle is especially powerful in reducing cognitive load, which can lead to improved user experience. According to a study published in *Interaction Design Foundation*, simplifying visual information through proximity can enhance comprehension and retention rates by up to 40%.

By organizing content in this way, you make it easier for users to understand the relationships between different pieces of information, reducing cognitive load and making the user experience more seamless.

Similarity: Unifying Design Elements

Another important principle is **similarity**, which states that elements that are visually similar are perceived as part of the same group. This can include similarities in color, shape, size, or font. For instance, using the same color scheme for all buttons that lead to a purchase can help users quickly identify actionable elements on your site. Similarly, consistent typography across headings, subheadings, and body text helps users differentiate between different levels of information.

A study by *MIT Sloan School of Management* found that websites using consistent design elements across pages had a 20% higher conversion rate than those that did not . By applying the principle of similarity, you create a cohesive design that guides users naturally through your content, enhancing both usability and aesthetic appeal.

Continuity: Guiding the User's Journey

Continuity is another Gestalt principle that plays a crucial role in design. It suggests that the human eye is inclined to follow lines, curves, or patterns in a continuous path.

This principle is particularly useful in guiding user navigation on a website or in an app.

For example, a curved line that leads from a product image to a call-to-action button can draw the user's eye along the intended path, making it more likely they will follow through with the action.

The continuity principle is backed by research from *EyeQuant*, which shows that designs that guide the user's gaze along a continuous path can increase engagement by up to 35%. Continuity helps create a smooth, uninterrupted experience that keeps users engaged and focused on the desired outcomes.

Closure: Encouraging Engagement Through Implied Completeness

The principle of **closure** involves the brain's ability to fill in missing parts of a visual image to perceive a complete, unified object. This principle can be used creatively in design to engage users by encouraging them to complete the image or message themselves.

For instance, a logo that uses negative space to suggest a shape or an icon can make the design more intriguing and memorable. Closure can also be applied in more subtle ways, such as in menu design, where partial visibility of additional options encourages users to explore further. A study in the *Journal of Vision* demonstrated that users are 25% more likely to remember incomplete shapes, logos, or designs due to the brain's natural inclination to seek closure. This can increase brand recall and engagement, making your content more memorable and impactful.

Figure-Ground: Emphasizing Key Elements

Fgure-ground is another Gestalt principle that is essential in design. This principle refers to the relationship between the foreground and background of an image. The human eye tends to separate objects from their surrounding background, which can be used to emphasize key elements in your design. For example, using a high-contrast background can make a call-to-action button stand out, drawing the user's attention to it immediately. Research by the *Interaction Design Foundation* indicates that effective use of figure-ground relationships can improve visual hierarchy and focus by 30% . By effectively using the figure-ground principle, you can control what elements stand out most in your design and ensure that users focus on the most important parts of your content.

Symmetry and Order: Creating a Balanced Experience

The principle of **symmetry and order**, also known as prägnanz, suggests that people perceive complex scenes in their simplest forms. In design, this means that users will naturally prefer and understand symmetrical and orderly layouts over chaotic or disorganized ones. A symmetrical design with balanced elements on either side of the page creates a sense of harmony and stability, which can make your website or app more pleasing to navigate.

According to a study in *The International Journal of Human-Computer Interaction*, symmetrical designs can reduce cognitive load by 30% and increase user satisfaction by 20% . This principle is particularly important in creating a professional and trustworthy brand image, as orderly designs tend to evoke feelings of reliability and clarity.

Common Fate: Reinforcing Grouping Through Movement

Another principle, **common fate**, refers to the perception that elements moving in the same direction are part of a single group. This principle can be used effectively in animation or interactive design. For example, if multiple elements move in unison when a user scrolls or clicks, it reinforces the idea that these elements are related and should be considered together.

This can be particularly useful in guiding users through a process or a series of steps, ensuring that they understand the relationship between different parts of your content. Research from *Cognitive Research: Principles and Implications* shows that common fate can improve task completion rates by up to 15% by visually reinforcing groupings.

Focal Points: Directing Attention Where It Matters

The principle of **focal points**, or emphasis, is crucial in ensuring that your design communicates its intended message clearly. Focal points are the areas of a design that naturally draw the eye first. By strategically placing and designing focal points, such as a prominent headline, a striking image, or a highlighted call-to-action button, you can guide the user's attention and control the flow of information.

A report from *ConversionXL* indicates that well-placed focal points can increase conversions by up to 60%, demonstrating their importance in guiding user behavior. This helps ensure that users see the most important parts of your message first and are more likely to engage with your content as intended.

Conclusion

Gestalt principles are not just abstract theories; they are practical tools that can greatly enhance the effectiveness of your design. By understanding how people naturally perceive visual information, you can create designs that not only look good but also communicate more effectively. As the renowned designer Dieter Rams once said, "Good design is as little design as possible."

Whether you're designing a website, an app, or a marketing campaign, applying Gestalt principles can help you craft a visual experience that resonates with your audience, makes your content easier to understand, and ultimately drives the desired action.

Call to Action (CTA)

Ready to optimize your design for better user perception and engagement? Start by applying Gestalt principles to your current digital assets and see how these small changes can lead to significant improvements in user experience and conversions.

Embrace the power of design psychology today and create visuals that truly connect with your audience!

~CHAPTER 17~

Loss Aversion

Tapping Into the Fear of Missing Out

Loss aversion is a powerful psychological principle suggesting that people are more motivated to avoid losses than to acquire equivalent gains. Research by *Daniel Kahneman* and *Amos Tversky*, who first introduced the concept, found that losses are psychologically twice as powerful as gains. In other words, the pain of losing something is often more intense than the pleasure of gaining something of equal value. This natural aversion to loss can be a compelling force in marketing, where tapping into the fear of missing out (FOMO) can drive customer decisions and increase conversions. By effectively leveraging loss aversion, you can create urgency and persuade your audience to take action before it's too late.

Creating Urgency with Limited-Time Offers

One of the most common ways to utilize loss aversion is through limited-time offers. When customers know that a discount or special offer is only available for a short period, they are more likely to act quickly to avoid missing out.

Studies show that 60% of millennials admit to making a reactive purchase because they were afraid of missing out on a limited-time offer. For example, if you're running a promotion on a survival gear kit, highlighting that the discount expires in 48 hours can create a sense of urgency. Customers who might otherwise have hesitated are driven by the fear of losing the opportunity to get a good deal.

The ticking clock plays on their natural aversion to loss, making them more likely to make a purchase sooner rather than later.

Leveraging Scarcity to Heighten Desire

Scarcity is another effective tool for leveraging loss aversion. When a product or service is presented as being in limited supply, it becomes more desirable because the potential for loss is heightened. If customers believe that a product is almost sold out or that there are only a few units left, they are more inclined to buy it immediately to avoid the disappointment of missing out.

According to *Robert Cialdini*, author of *Influence: The Psychology of Persuasion*, scarcity is one of the most effective principles in influencing human behavior. For instance, if you're selling a popular weight loss supplement, displaying a message like "Only 5 left in stock" can trigger a rush to purchase, as customers don't want to lose the chance to get the product before it's gone.

Using Free Trials and Introductory Offers to Minimize Risk

Free trials and introductory offers can also be designed to tap into loss aversion. Offering a free trial for a subscription service, for example, allows customers to experience the benefits without any initial commitment. However, as the trial period comes to an end, reminding users of what they will lose if they don't continue the subscription can be a powerful motivator.

A study by *Harvard Business School* found that customers are more likely to convert to paid plans when reminded of the benefits they will lose after a free trial ends.

The thought of losing access to a service they've begun to enjoy can push them to convert from a free trial to a paid subscription. This approach turns the potential loss of benefits into a compelling reason to commit.

Framing Messages Around Potential Losses

Another way to invoke loss aversion is by framing your messaging around what customers stand to lose rather than what they could gain. For example, instead of saying, "Save $100 on this course," you might frame it as, "Don't miss out on saving $100 by enrolling today." Research published in the *Journal of Economic Behavior & Organization* indicates that loss-framed messages are significantly more persuasive than gain-framed messages, especially in contexts where immediate action is desired.

The second phrasing emphasizes the potential loss of savings, which can be more persuasive. This subtle shift in language plays on the psychological discomfort people feel when faced with the possibility of losing something, even if it's something they haven't yet acquired.

Social Proof and FOMO

Social proof can also be used to leverage loss aversion. When customers see that others are purchasing or enjoying a product, they may feel that if they don't act, they will miss out on an experience that others are benefiting from. This is particularly effective in the context of popular products or services that have a strong following. Highlighting customer testimonials, reviews, or the number of people who have already taken advantage of an offer can create a sense of community and urgency, encouraging others to join in to avoid being left out.

A survey by *Nielsen* found that 92% of consumers trust recommendations from friends and family over any other form of advertising, underscoring the power of social proof in creating a sense of urgency and fear of missing out.

Early Bird Pricing to Encourage Quick Decisions

Early bird pricing is another tactic that capitalizes on loss aversion. Offering a lower price for those who act early, and then gradually increasing the price as the deadline approaches, creates a strong incentive to buy sooner than later. Customers who delay their decision risk paying more, which taps into their fear of losing the opportunity to save money. This strategy can be particularly effective for events, courses, or product launches, where the anticipation of a price increase can drive early sales. Data from *Eventbrite* shows that 24% of tickets are sold during the early bird phase, reflecting the effectiveness of this strategy in driving initial sales.

Structuring Loyalty Programs to Encourage Ongoing Engagement

Membership or loyalty programs can also be structured to take advantage of loss aversion. For example, if customers know that their points or rewards will expire if not used within a certain timeframe, they are more likely to take action to redeem them. The idea of losing something they've earned—whether it's a discount, a free item, or an exclusive perk—can prompt customers to make additional purchases or engage more frequently with your brand.

A study by *Forrester Research* found that 72% of consumers who participate in loyalty programs feel more engaged and connected with the brand, showing how fear of losing rewards can enhance customer loyalty and increase repeat purchases.

Framing Refund Policies to Reduce Perceived Risk

Another application of loss aversion is in the framing of refund policies. Offering a money-back guarantee can reduce the perceived risk of making a purchase, but highlighting the fact that the guarantee is only available for a limited time can create a sense of urgency. For example, a message like "Get a full refund within 30 days if you're not satisfied" reassures customers, while also subtly reminding them that the opportunity to get their money back won't last forever. This taps into their desire to avoid the loss of money or satisfaction, making them more likely to act within the given timeframe.

Exclusive Content and Limited Access

Loss aversion is not just about discounts and deals; it can also be applied to content and information. For instance, offering exclusive content that is only available for a limited time or to a select group of subscribers can create a fear of missing out. If customers know that they will lose access to valuable information if they don't act quickly, they are more likely to engage with the content, sign up for a newsletter, or share their contact information. A study by *Digital Marketing Institute* found that emails offering exclusive, time-sensitive content had a 20% higher open rate than standard promotional emails, demonstrating the power of FOMO in driving engagement.

Conclusion

By effectively leveraging loss aversion, you can tap into a powerful psychological motivator that drives people to act. As Daniel Kahneman noted, "We are driven more strongly to avoid losses than to achieve gains." The fear of missing out or losing something valuable can be more compelling than the promise of gain.

When you frame your offers, products, and messaging around the potential for loss, you create a sense of urgency and importance that can significantly boost conversions.

Whether through limited-time offers, scarcity tactics, or exclusive content, invoking loss aversion can help you create more persuasive marketing strategies that resonate deeply with your audience and encourage them to take action.

Call to Action

Ready to harness the power of loss aversion in your marketing strategy?

Start by evaluating your current campaigns to identify opportunities to create urgency and highlight potential losses. Use these tactics to boost engagement and conversions, ensuring that your audience feels compelled to act quickly and decisively. Implement loss aversion techniques today and watch your conversions soar!

~CHAPTER 18~

Novelty

Capturing Attention with the New and Unique

Novelty is a powerful driver of human behavior. People are naturally drawn to things that are new, unique, and different. In a world saturated with information and countless options, capturing attention requires standing out from the crowd. Novelty not only grabs attention but also sparks curiosity, making people more likely to explore, engage, and ultimately convert. By leveraging the appeal of the new and unique, you can create marketing strategies that resonate with your audience and drive results.

Why Novelty Works

The first step in using novelty effectively is to understand why it works. The human brain is wired to seek out new experiences and information. Research from the *Journal of Neuroscience* reveals that exposure to novel stimuli activates the brain's reward system, releasing dopamine, a neurotransmitter associated with pleasure and motivation.

When something unfamiliar catches our eye, it activates the brain's reward centers, making us feel excited and eager to learn more. This is why new products, fresh ideas, or unique approaches can be so compelling—they offer the promise of something different, breaking the monotony and stimulating curiosity. In marketing, tapping into this desire for novelty can set your brand apart and make it more memorable.

Product Launches: Generating Buzz with New Offerings

One of the most effective ways to harness novelty is through product launches. Introducing a new product or service is a natural opportunity to capture attention. The excitement of something new can drive interest and create buzz around your brand. A study by *Harvard Business Review* found that new product launches can boost sales by up to 25% in the first quarter alone if effectively marketed.

For example, if you're in the survival niche, launching a new piece of gear with innovative features can attract both existing customers and new prospects. The key is to highlight what makes the product unique—whether it's a breakthrough in technology, a design that solves a common problem, or a feature that sets it apart from anything else on the market. By focusing on what's new and different, you can generate interest and drive sales.

Limited Editions: Creating Urgency with Scarcity and Novelty

Another way to incorporate novelty is through limited-edition offerings. Scarcity combined with the appeal of something new can create a powerful sense of urgency. Customers who might hesitate to purchase a regular item may be more inclined to buy if they know it's a limited edition or available for a short time.

Research from the *Journal of Retailing* shows that scarcity marketing, combined with novelty, can increase perceived value by 40% and boost purchase intent significantly. This strategy works well for products like seasonal items, special collaborations, or exclusive designs.

For example, if you're selling fitness apparel, releasing a limited-edition color or pattern can attract attention and motivate customers to act quickly to avoid missing out on something unique.

Innovative Campaigns: Breaking the Mold to Engage Audiences

Novelty can also be applied to marketing campaigns themselves. A fresh, unexpected approach to your messaging or branding can reinvigorate your audience's interest in your brand. Innovative campaigns that utilize new formats, such as interactive content, augmented reality, or gamified experiences, can make your marketing efforts stand out. According to a study by *Adobe Digital Insights*, interactive and novel content formats increase user engagement rates by 30% on average.

For instance, launching a campaign that uses a new format, such as interactive content, augmented reality, or a gamified experience, can make your marketing efforts stand out. This not only captures attention but also encourages deeper engagement as customers explore the new format. In the weight loss industry, for example, creating a virtual challenge that uses a new app or technology can attract participants who are looking for a fresh way to achieve their goals.

Fresh Content: Keeping Your Audience Engaged

Innovative content is another way to leverage novelty. Regularly introducing new types of content, such as video series, podcasts, or live streams, can keep your audience engaged and coming back for more.

People appreciate brands that continue to offer fresh perspectives and insights, rather than recycling the same old content.

A study by *HubSpot* found that brands that regularly update their content see 55% higher visitor engagement rates than those that do not. If you're in the marketing niche, for example, starting a new podcast that interviews industry leaders or explores cutting-edge strategies can position your brand as a thought leader and keep your content offerings dynamic and exciting.

Small Changes, Big Impact: Keeping Your Digital Presence Fresh

Even small changes can have a big impact when it comes to novelty. Regularly updating your website's design, introducing new features, or refreshing your branding can keep your digital presence feeling current and engaging. A website that always looks the same can start to feel stale, but periodic updates can make it feel like there's always something new to discover. According to *Google's UX Playbook for Retail*, 75% of users judge a website's credibility based on its design. This doesn't mean overhauling your entire site every few months, but rather making thoughtful updates that reflect new trends, customer feedback, or your evolving brand story.

Tailoring Novelty to Your Audience

Novelty isn't just about what's new in the market—it's also about what's new to the customer. Tailoring your marketing to introduce customers to something they haven't seen before, even if it's a product or service you've offered for a while, can reignite interest. This might involve highlighting a lesser-known feature, offering a new use case, or rebranding an existing product to target a different audience. By presenting your offerings in a new light, you can capture the attention of customers who might have overlooked them previously.

Collaborations and Partnerships: Combining Forces for Something Unique

Collaborations and partnerships are another way to introduce novelty. By teaming up with another brand, especially one from a different industry or with a different audience, you can create something truly unique that neither brand could offer on its own.

A report by *MarketingProfs* found that co-branded campaigns are 25% more effective than solo campaigns in driving new customer acquisition.

For example, if you're in the survival gear industry, partnering with a well-known outdoor adventure brand to create a co-branded product line can attract attention from both audiences and create a buzz that drives sales. Collaborations bring together different strengths and perspectives, resulting in a product or campaign that feels fresh and exciting.

Refreshing Communication: Keeping Your Brand Dynamic

Novelty can also be leveraged in how you communicate with your audience. Introducing new messaging, slogans, or visual styles in your advertising can refresh your brand's image and attract attention. If your brand has been using the same tagline or imagery for years, trying something new can reinvigorate your marketing and appeal to a broader audience. This doesn't mean abandoning your brand identity, but rather evolving it in ways that keep it relevant and engaging.

Small Surprises: Delighting Customers with the Unexpected

Novelty doesn't always have to be about big changes. Even small, unexpected touches can make a lasting impression. Personalizing your communications with customers, sending them surprise gifts, or offering exclusive sneak peeks of upcoming products are all ways to introduce a sense of novelty into the customer experience.

These little surprises can delight your customers, making them feel valued and keeping them engaged with your brand. A study by *Bain & Company* found that customers who receive personalized experiences are 40% more likely to become repeat customers.

Conclusion

Incorporating novelty into your marketing strategy is about more than just being different for the sake of it. It's about understanding the power of the new and unique to capture attention, generate excitement, and motivate action. Whether through product launches, limited editions, innovative campaigns, or fresh content, leveraging novelty can help your brand stand out in a crowded marketplace. As Albert Einstein famously said, *"Creativity is intelligence having fun."* Embrace creativity in your marketing and watch your audience engagement and conversions soar.

Call to Action

Ready to infuse your marketing strategy with the power of novelty? Start by brainstorming fresh ideas and innovative approaches that can set your brand apart.

Experiment with new content formats, launch a unique product, or partner with another brand for an exciting collaboration.

Take the first step today and discover the impact of novelty on your audience's engagement and your bottom line!

~CHAPTER 19~

Pricing

Strategies to Shape Perceptions of Value

Pricing is one of the most powerful tools in marketing because it directly influences how customers perceive the value of your product or service. The right pricing strategy can make your product seem like a great deal, a luxury item, or anything in between. It's not just about what you charge—it's about how you present that charge in a way that aligns with the value customers place on your offering. By carefully crafting your pricing strategy, you can shape perceptions, drive sales, and build long-term customer relationships.

Psychological Pricing: The Subtle Art of Perception

One of the most effective pricing strategies is psychological pricing, which involves setting prices that have a psychological impact on the customer. For example, pricing something at $9.99 instead of $10.00 takes advantage of the left-digit effect, where customers perceive the price as being closer to $9 rather than $10 buy. It's a subtle tactic, but it plays a big role in how customers assess value.

According to a study published in the *Journal of Consumer Research*, prices ending in .99 can increase sales by 24% compared to rounded numbers because they create an illusion of a bargain without substantially lowering the actual price.

Anchoring: Setting the Stage for Perceived Value

Anchoring is another powerful pricing strategy. It involves presenting a higher-priced item first to set a reference point, or "anchor," in the customer's mind. Once this anchor is established, other prices are viewed in relation to it.

For instance, if you introduce a high-end product at $500 and then show a more affordable option at $300, the second product seems like a much better deal simply because it's being compared to the higher price. A classic experiment by Dan Ariely, a behavioral economist, showed that when consumers are presented with an initial higher price, they tend to spend more, as subsequent prices feel like bargains. This strategy can make your mid-range products more appealing and increase the likelihood of a sale.

Bundling: Creating Value Through Packages

Bundling is a pricing strategy that enhances perceived value by offering multiple products or services together at a reduced rate. Customers often feel they're getting more for their money when they purchase a bundle, even if they don't need every item included.

For example, if you're selling fitness equipment, bundling a yoga mat, resistance bands, and a workout guide together for a single price can make customers feel they're getting a comprehensive package at a better value than if they bought each item separately. A report from *Harvard Business Review* suggests that bundling can increase sales by up to 30% because it simplifies the decision-making process and reduces price sensitivity.

Bundling also encourages customers to spend more than they might have on a single product, increasing your average order value.

Tiered Pricing: Offering Choices to Fit Every Budget

Tiered pricing offers different versions of a product at varying price points, allowing you to capture a wider range of customers by providing options that cater to different budgets and needs. For example, offering a basic, standard, and premium version of a software product allows customers to choose the level of service that best fits their needs. The existence of a premium option can also make the mid-tier option appear more valuable, as it's often seen as a compromise between affordability and features. Research by *McKinsey & Company* found that companies using tiered pricing strategies saw an average revenue increase of 20% because it appeals to diverse customer segments and maximizes conversion rates.

Decoy Pricing: Guiding Choices with a Strategic Option

Price anchoring can be further enhanced by introducing a decoy pricing option. This involves adding a third option that is priced close to your preferred product but offers less value, making the intended choice seem more appealing. For example, if you have a standard product at $100, a premium version at $150, and a decoy version at $140 that offers fewer features than the premium, customers are more likely to see the $150 option as the best value. This strategy subtly guides customers toward the product you want to sell the most.

Research by *The Economist* on their own pricing strategy demonstrated that adding a decoy option increased subscriptions by 50%, proving the effectiveness of this tactic in influencing customer behavior.

Discounts and Promotions: Strategic Use of Incentives

Offering discounts and promotions can also shape how customers perceive your pricing and value. However, it's important to use discounts strategically to avoid devaluing your product. Flash sales, limited-time offers, and exclusive discounts for loyal customers can create urgency and incentivize purchases without permanently lowering your prices.

A study by *RetailMeNot* found that 80% of consumers feel motivated to make a first-time purchase with a brand that is new to them if they find an offer or discount. For example, offering a 20% discount on a popular item for 48 hours can drive quick sales by playing on customers' fear of missing out. This not only boosts sales in the short term but can also enhance the perceived value of the product when it returns to its regular price.

Price Framing: Highlighting Value Through Comparisons

Price framing involves presenting the price relative to other factors, such as comparing the cost of a product to the potential savings or benefits it offers.

For example, if you're selling a home security system, you might frame the price by highlighting the cost of potential damages from a break-in that the system could prevent. This comparison makes the price seem more reasonable by focusing on what the customer stands to gain or save in the long run.

A *Stanford University* study found that framing pricing in terms of loss aversion rather than gain can increase conversion rates by 30%, as people are more motivated to avoid losses than to acquire equivalent gains.

Subscription Pricing: Building Long-Term Relationships

Subscription pricing offers customers access to a product or service for a recurring fee, often monthly or annually. The appeal of subscription pricing lies in its predictability and perceived affordability, as customers are paying smaller amounts over time rather than a large upfront cost.

According to *Zuora's Subscription Economy Index*, companies with a subscription-based model grow revenues 5 times faster than S&P 500 companies, highlighting the effectiveness of this strategy.

For example, a subscription to an online fitness program may seem more accessible at $19.99 per month rather than a one-time payment of $240. This model can also build long-term customer relationships, as the recurring nature of the payment keeps customers engaged with your product over a longer period.

Price Localization and Dynamic Pricing: Tailoring Prices for Maximum Impact

Price localization involves adjusting your prices based on geographic location, currency, or local market conditions. This approach can help make your products more accessible to a global audience by considering the purchasing power and competition in different regions. For example, pricing a product differently in emerging markets compared to more developed markets allows you to maximize sales while maintaining a fair price point that aligns with local expectations.

Dynamic pricing is a flexible strategy where prices change based on demand, competition, or other external factors. This approach, often used in industries like travel and hospitality, allows you to optimize pricing for maximum profitability while still offering competitive rates to your customers.

A study by *MIT Sloan School of Management* found that dynamic pricing can increase revenues by up to 10% by allowing companies to adjust prices in real-time to reflect current market conditions and consumer demand.

Conclusion

Pricing is not just a number; it's a strategic tool that can significantly impact how customers perceive the value of your product or service. By employing psychological pricing, anchoring, bundling, tiered pricing, and other strategies, you can influence customer decisions and enhance the perceived value of what you offer.

The key is to align your pricing strategy with your overall brand positioning and customer expectations, ensuring that your prices not only attract buyers but also reinforce the value of your brand. When done correctly, pricing becomes a powerful lever for driving sales, increasing customer satisfaction, and building long-term loyalty.

Call to Action

Ready to reshape your pricing strategy for maximum impact?

Start by evaluating your current approach and identify areas where you can implement these powerful techniques.

Experiment with different strategies, test their effectiveness, and refine your approach to better align with your audience's perceptions of value.

Take action today to optimize your pricing and watch your conversions soar!

~CHAPTER 20~

Priming

Preparing the Mind for Persuasion

Priming is a psychological technique that involves subtly influencing how people think and feel before they make a decision. By exposing your audience to certain stimuli, you can shape their perceptions, attitudes, and behaviors in a way that prepares them for persuasion. In marketing, priming can be a powerful tool to guide your customers toward a desired action by creating a mental context that makes your message more effective.

The Science Behind Priming

Priming works because the human brain is constantly processing and interpreting information, often subconsciously. When exposed to certain cues or stimuli, the brain becomes more receptive to related ideas or actions. Research by psychologist John Bargh at Yale University demonstrated how priming could influence behavior. In one experiment, participants who were subconsciously primed with words related to old age walked more slowly than those who were not primed. This shows how subtle cues can impact actions, even when people are unaware of their influence.

In marketing, using similar techniques can help prime your audience to be more receptive to your messages and offers.

Visual Cues: Setting the Stage for Perception

One of the most common ways to use priming in marketing is through visual cues. Colors, images, and design elements can all be used to set the tone and influence how customers perceive your brand and products.

For example, if you're selling luxury goods, using colors like gold, black, and deep blue can prime customers to think of exclusivity and high value. This makes them more receptive to premium pricing and positions your products as worth the investment. Similarly, in the health and wellness industry, using clean, bright colors like green and white can prime customers to associate your products with purity and well-being, making them more likely to trust your offerings. Studies have shown that color can increase brand recognition by up to 80%, highlighting the importance of visual priming in marketing.

Language: Words That Prime for Action

Language is another powerful tool for priming. The words you choose in your headlines, product descriptions, and calls to action can set the stage for how customers perceive your message. Words like "imagine," "discover," or "unlock" can prime customers to think about the potential benefits of your product, making them more open to persuasion. A study published in the *Journal of Personality and Social Psychology* found that priming people with action-oriented words increased their likelihood of taking action, such as participating in a study or completing a task.

If you're marketing a weight loss program, starting with a phrase like "Imagine yourself achieving your goals effortlessly" can create a mental image that primes the customer to believe in the possibility of success, making them more likely to engage with your offer.

Social Proof: Creating Expectations Through Testimonials

Priming can also be achieved through the use of testimonials and social proof. When customers see others who have had positive experiences with your product, it primes them to expect similar results. This is particularly effective when the testimonials come from people who are similar to your target audience. According to a study by Nielsen, 92% of consumers trust recommendations from people they know, and 70% trust consumer opinions posted online, making social proof a powerful priming tool.

For example, if you're selling fitness equipment, showcasing testimonials from everyday people who have achieved their fitness goals with your products can prime new customers to believe that they too can succeed. This technique works by creating a sense of relatability and trust before the customer even considers making a purchase.

Strategic Information Placement: Guiding Perception Through Order

Another way to use priming is through the strategic placement of information. The order in which information is presented can influence how it's perceived and remembered. Leading with a strong benefit or a compelling statistic can prime customers to view the rest of your message in a positive light. A study by *The Journal of Marketing* found that information presented first is more likely to be remembered and can shape the perception of subsequent information.

If you're promoting a financial planning service, starting with a powerful statement like "89% of our clients achieve their retirement goals" primes customers to think that your service is effective, setting the stage for the rest of your pitch. This approach uses the power of first impressions to shape how subsequent information is interpreted.

Repetition and Consistency: Reinforcing Brand Associations

Priming can also occur through repetition. By consistently reinforcing certain messages or themes, you can prime your audience to associate those ideas with your brand. For example, if you repeatedly emphasize quality and craftsmanship in your marketing materials, customers are more likely to associate your brand with those qualities, even before they experience your product firsthand. This type of priming can create strong brand associations that influence customer perceptions and buying decisions over time.

Environmental and Timing Priming: Contextual Cues That Influence Behavior

Environmental priming is another technique that can be used in physical retail spaces or on websites. The environment in which a customer encounters your product can influence their mood and perception. For example, playing soft, relaxing music in a spa can prime customers to feel calm and pampered, making them more receptive to purchasing luxury treatments. On a website, using clean, organized layouts can prime users to feel that your brand is professional and trustworthy, encouraging them to stay longer and explore more of your offerings.

Timing also plays a crucial role in priming. The context in which a message is delivered can affect how it's received. Sending an email campaign about holiday discounts in the weeks leading up to Christmas primes customers to think about gift-giving and shopping, making them more likely to respond positively to your offers. Research published in the *Journal of Consumer Research* shows that messages aligned with temporal context are more likely to be effective, as they resonate with customers' current state of mind and needs.

Content Priming: Preparing the Mind Through Relevant Information

Content priming involves using related content to prepare the customer for your main message. For instance, if you're promoting a new skincare product, a blog post about the importance of skincare routines can prime readers to be more interested in your product. By educating and engaging the customer with relevant content first, you create a mental framework that makes them more receptive to your product pitch. This approach not only primes the customer for persuasion but also builds trust by providing valuable information upfront.

Conclusion

Priming isn't about manipulation; it's about setting the stage for effective communication. By understanding how people process information and make decisions, you can create marketing strategies that resonate more deeply with your audience. Whether through visual cues, language, repetition, or timing, priming prepares the mind for persuasion, making your messages more impactful and your calls to action more compelling. When done thoughtfully, priming can enhance the effectiveness of your marketing efforts, leading to better customer engagement, higher conversions, and stronger brand loyalty.

Call to Action

Ready to apply the power of priming to your marketing strategy? Start by identifying the key messages you want to reinforce and consider how you can use visual cues, language, and timing to set the stage for persuasion.

Test different priming techniques, measure their impact, and refine your approach to maximize your results.

Begin today to influence your audience more effectively and drive your business forward!

~CHAPTER 21~

Eye Tracking & Eye Gaze

Optimizing Layouts for Maximum Impact

Eye tracking and eye gaze are essential concepts in design and marketing that focus on how people visually engage with content. By understanding where and how your audience's eyes move across a page, you can optimize your layouts to ensure that the most important information is seen first and that your call to action is prominently highlighted.

This approach helps create a more effective user experience, guiding visitors naturally through your content and increasing the likelihood of conversion. The primary goal of eye tracking is to determine which elements on a page draw the most attention and how the viewer's gaze progresses from one element to another. Studies have shown that people generally follow predictable patterns when viewing web pages, and understanding these patterns can help you design layouts that are both engaging and-effective.

Understanding Eye Tracking Patterns

One of the most common eye-tracking patterns is the **F-pattern**. This pattern shows that users tend to start by scanning horizontally across the top of the page, then move down slightly and scan horizontally again, and finally, move vertically down the left side of the page. This behavior resembles the shape of the letter "F.

Knowing this, you can optimize your design by placing the most important information, such as your headline, key benefits, and calls to action, along the top and left side of the page. This ensures that these elements are seen quickly and are more likely to capture the viewer's attention.

Another important pattern is the **Z-pattern**, which is often seen on simpler, less text-heavy pages. In this pattern, viewers start at the top left corner, move horizontally across to the top right, then diagonally down to the bottom left, and finally across to the bottom right. This Z-shaped movement is common on pages with a strong visual hierarchy and clear, bold elements like images or headlines.

To take advantage of this pattern, you can place your logo or main message in the top left, supporting images or information in the middle, and your call to action in the bottom right. This layout guides the viewer's eyes naturally through the content, making it more likely that they'll reach the desired action.

Leveraging Visual Elements for Maximum Engagement

Eye gaze direction plays a significant role in how users interact with a page. People's eyes are naturally drawn to faces, especially eyes, which means that using images of people looking directly at the camera can capture attention quickly. However, you can also use this to your advantage by having the subject in the image look towards your call to action or another key element on the page.

A study published in the *Journal of Consumer Research* found that images where the subject's gaze directed toward a product or call to action increased the likelihood of user engagement by over 50%. This subtle cue can direct the viewer's gaze toward the most important part of your content, increasing the likelihood that they'll notice and interact with it.

Contrast and color are also critical in directing eye gaze. High-contrast elements stand out more and are likely to draw attention first. For example, a brightly colored button on a page with a muted background will immediately catch the viewer's eye, making it an effective way to highlight a call to action.

Similarly, using contrasting colors for important text, such as headlines or key benefits, can ensure these elements are seen quickly. Research shows that the use of contrasting colors can increase the visibility of key elements by up to 60%, emphasizing the importance of strategic color choices.

The Role of White Space and Typography

White space, or negative space, is another crucial factor in optimizing layouts for eye tracking. While it might seem like empty space, white space actually helps to focus attention on the content that matters. By giving elements room to breathe, you can prevent the page from feeling cluttered and guide the viewer's eyes more effectively.

For example, surrounding a call-to-action button with white space can make it stand out more, drawing attention and increasing the chances of a click. Proper use of white space creates a cleaner, more navigable design that keeps the viewer's attention where you want it.

Typography also influences eye movement. Larger fonts and bold text naturally draw the eye, making them ideal for headlines and important information.

However, consistency in font size and style across similar elements helps guide the viewer smoothly through the content.

For example, maintaining a consistent font size for all body text allows readers to focus on the message rather than being distracted by changes in style. Meanwhile, using a distinct, larger font for headings helps to break up content into digestible sections, making it easier to follow the flow of information.

Optimizing Interactive Elements and Visual Hierarchy

Placement of **interactive elements**, such as buttons and links, should be informed by eye-tracking data as well. Placing these elements where the eye naturally travels increases the likelihood of engagement. For instance, placing a call-to-action button directly after a key benefit statement or in the path of a Z-pattern can encourage clicks because it's positioned where the viewer's eye is already going.

Additionally, ensuring that these elements are visually distinct and easy to find helps reduce friction in the user experience, leading to higher conversion rates.

Visual hierarchy involves arranging elements on the page in order of importance, with the most critical elements standing out the most. Eye tracking shows that users are more likely to engage with content that is organized in a clear and logical manner.

For example, placing a large, bold headline at the top of the page, followed by smaller subheadings and then body text, helps guide the viewer through the content in a way that makes sense. By establishing a strong visual hierarchy, you ensure that viewers absorb the most important information first, setting the stage for effective communication.

Adapting for Mobile Optimization

Mobile optimization is essential in today's digital landscape. Eye-tracking patterns differ slightly on mobile devices, where screens are smaller and scrolling is more common. On mobile, users often scan vertically down the center of the screen, so placing key elements like headlines, images, and calls to action in the middle of the page can be more effective.

Ensuring that your design is responsive and adapts to different screen sizes will help maintain the effectiveness of your layout, no matter how your audience accesses your content. A study by Google found that mobile-friendly sites increase conversion rates by up to 67%, underscoring the importance of optimizing for mobile.

Conclusion

By leveraging eye tracking and eye gaze principles, you can create layouts that not only capture attention but also guide your audience through your content in a way that maximizes impact.

Whether through strategic placement, use of color and contrast, or effective visual hierarchy, understanding how people view and interact with your page allows you to design with intention. This leads to better engagement, higher conversion rates, and ultimately, more successful marketing efforts.

Call to Action

Ready to optimize your website for maximum impact?

Start implementing eye-tracking strategies today to enhance user experience and boost conversions.

Evaluate your current layout, identify key areas for improvement, and use these insights to create a more engaging and effective online presence. Take the next step towards increasing your conversions now!

~CHAPTER 22~

Understanding Emotions

The Key to Connecting with Your Audience

Understanding emotions is crucial in marketing because emotions drive decisions. When you tap into your audience's feelings, you create a connection that goes beyond the product or service you're offering. This connection makes your brand memorable and influences how people perceive and interact with what you offer. It's not just about selling a product; it's about resonating with your audience on an emotional level, which can lead to deeper loyalty and higher conversion rates.

The Role of Emotions in Decision-Making

Emotions like joy, fear, trust, and excitement significantly impact how people react to your marketing messages. Research from the *Journal of Consumer Research* suggests that emotional responses to ads are more influential on a person's intent to buy than the ad's content itself. This is because emotions create a strong, immediate response that logic alone cannot achieve.

As the psychologist Daniel Kahneman noted, "We think much less than we think we think," highlighting that emotions often guide our decisions more than rational thought. **Joy** is a powerful emotion in marketing.

When you create content that makes your audience feel happy, they're more likely to associate those positive feelings with your brand.

This could be through heartwarming stories, humor, or content that showcases the joy your product brings. For example, Coca-Cola's "Open Happiness" campaign effectively uses joy to create a positive emotional association with the brand.

In your own marketing, consider how you can use joy to make your audience feel good about their interactions with your brand.

Fear can also be a potent motivator, especially when it creates a sense of urgency. Fear-based marketing often revolves around the idea of avoiding a negative outcome. In the survival niche, for instance, emphasizing the fear of being unprepared in an emergency can compel customers to purchase your products to mitigate that fear. The key is to use fear responsibly, ensuring that it's balanced with a clear solution your product or service provides. A study from *Psychology & Marketing* found that fear appeals in marketing can be highly effective when combined with a specific, actionable solution.

Trust is another critical emotion that plays a vital role in connecting with your audience. People are more likely to buy from brands they trust. Building trust involves consistency, transparency, and delivering on promises. For example, providing high-quality products and exceptional customer service can build a reputation that fosters trust. Testimonials and reviews are also powerful tools in establishing trust, as they show potential customers that others have had positive experiences with your brand.

A 2020 survey by Edelman found that 81% of consumers need to trust a brand to consider buying from them. When people trust your brand, they're more likely to become repeat customers and recommend you to others, reinforcing the importance of trust in driving customer loyalty.

Excitement can energize your audience and create anticipation around your products or services. Launching a new product with a teaser campaign that builds excitement can generate buzz and drive interest. For example, in the fitness industry, introducing a new workout program with videos showcasing dynamic results can excite your audience about the potential benefits, leading to increased sign-ups or purchases. This excitement creates a strong emotional connection that can boost your brand's visibility and appeal.

Empathy involves putting yourself in your customers' shoes and addressing their pain points in a way that shows you care. For example, in the weight loss industry, acknowledging the challenges of maintaining a healthy lifestyle and offering support through your products or services can resonate deeply with your audience.

According to a study published in the *Harvard Business Review*, brands that show empathy and understand their customers' emotions are more likely to develop long-term loyalty. This emotional connection can make your brand stand out as one that genuinely cares about its customers' well-being.

Surprise is another emotion that can be used to delight your audience and make your brand more memorable. Unexpected rewards, personalized offers, or a unique unboxing experience can create moments of surprise that leave a lasting impression. For example, sending a thank-you gift to loyal customers can create a sense of delight that strengthens their emotional connection to your brand. These small but impactful gestures can turn satisfied customers into enthusiastic brand advocates.

Sadness, when used carefully, can also create a strong emotional impact. While it's important not to exploit negative emotions, sharing stories that evoke empathy, and compassion can humanize your brand and build deeper connections. For example, if your brand is involved in charitable work, sharing the stories of those you've helped can touch your audience's hearts and inspire them to support your cause. This approach can lead to stronger customer loyalty as people feel they're part of something meaningful by supporting your brand.

Leveraging Emotions for Deeper Connections

Understanding the emotional triggers of your audience allows you to craft messages that resonate on a deeper level. It's not just about identifying the right emotions but also about knowing when and how to use them effectively in your marketing strategies. For instance, **fear and urgency** might be more effective in the final stages of a sales funnel, while **trust and empathy** are crucial during the initial stages of brand awareness and engagement. Building an emotional connection requires consistency and authenticity. Customers can quickly sense when a brand is being disingenuous. Therefore, it's crucial to align your messaging, visuals, and overall brand ethos with the emotions you want to evoke. When done correctly, this strategy not only enhances customer loyalty but also encourages word-of-mouth marketing, as emotionally connected customers are more likely to share their positive experiences with others.

Conclusion

Understanding and leveraging emotions in marketing is more than a tactic; it's a strategy for building lasting connections with your audience. By tapping into the right emotions—whether it's joy, fear, trust, or empathy—you can create a powerful bond with your customers that goes beyond the transactional. This emotional engagement can lead to higher conversions, stronger loyalty, and greater advocacy for your brand.

Call to Action

Ready to connect with your audience on a deeper emotional level?

Start by identifying the key emotions that resonate with your target market and integrate them into your marketing strategy. Create campaigns that not only inform but also inspire, delight, and engage, turning casual customers into loyal fans.

Begin today and watch your customer relationships grow stronger and more meaningful.

~CHAPTER 23~

Creating More Effective Ads

Techniques to Captivate and Convert

Creating more effective ads is essential for capturing attention and driving conversions in a crowded marketplace. With countless messages competing for your audience's focus, your ads need to stand out not just visually, but emotionally and intellectually. The most effective ads don't just promote a product; they tell a story, evoke emotions, and persuade the viewer that taking action is the best choice.

Understanding Your Audience

One of the most important techniques for creating effective ads is understanding your audience. Knowing who you're speaking to allows you to tailor your message to their needs, desires, and pain points. A survey by *HubSpot* revealed that ads tailored to the audience's preferences are twice as effective in driving engagement as generic ones.

For example, if you're targeting busy professionals with a fitness product, your ad should highlight how the product saves time and fits into a hectic schedule, rather than just focusing on its features.

This approach ensures that your message resonates with the audience and makes them feel that your product is specifically designed to meet their needs.

Clarity and Simplicity

Clarity is key in ad creation. Your message needs to be clear and straightforward. Overloading an ad with too much information or trying to communicate multiple messages at once can confuse the viewer and dilute the impact.

Research from the *Journal of Advertising Research* shows that ads with a single, clear message have a 45% higher recall rate than those with multiple messages. A single, strong message that speaks directly to your audience's needs or desires is far more effective. For instance, an ad for a survival kit should focus on the key benefit—being prepared for any emergency—rather than listing every item in the kit. This concise messaging cuts through the noise and makes the ad more memorable.

Visual Appeal

Visual appeal is another crucial element in creating effective ads. Strong visuals grab attention and communicate your message quickly. A study from *MIT* found that people can process and understand images in as little as 13 milliseconds, highlighting the importance of strong visual elements in ads. The use of color, imagery, and layout should all work together to draw the viewer in and direct their focus to the most important parts of the ad.

For example, a bold, high-contrast color scheme can make a product stand out, while a well-placed image can evoke the right emotion or context. In the weight loss industry, showing before-and-after images can be particularly powerful, as they visually demonstrate the effectiveness of the product or program, making the benefits immediately clear.

Storytelling and Emotion

Storytelling is a powerful technique in ad creation. A good story can engage the viewer on an emotional level, making your message more relatable and compelling. According to *Nielsen*, ads that tell a story generate a 70% increase in effectiveness compared to those that don't. Instead of just presenting facts, tell a story that illustrates how your product or service can change someone's life.

For example, an ad for a financial planning service might tell the story of a family that achieves financial security and peace of mind through careful planning. This narrative approach helps the viewer see themselves in the story, making the benefits of the service feel more tangible and attainable.

Emotion plays a significant role in ad effectiveness. Ads that evoke strong emotions—whether it's happiness, fear, excitement, or nostalgia—are more likely to be remembered and acted upon. A study from *The Harvard Business Review* found that ads that evoke emotions have twice the effectiveness of those that focus on rational content.

Emotional appeals can create a deeper connection with the viewer, making them more likely to engage with the ad and take the desired action. For example, an ad for a home security system might evoke the fear of an unsafe home environment, but then immediately provide the solution by showcasing the peace of mind that comes with using the product. This emotional journey can be highly persuasive.

Social Proof and Call-to-Action

Social proof is another technique that can boost the effectiveness of your ads. Featuring testimonials, reviews, or endorsements from satisfied customers can build trust and credibility.

According to a survey by *BrightLocal*, 91% of consumers trust online reviews as much as personal recommendations. People are more likely to trust a product if they see that others have had positive experiences with it. For example, an ad for a skincare product might include a brief testimonial from a real customer who achieved great results, along with before-and-after photos. This kind of social proof can reassure potential customers that your product delivers on its promises, increasing their likelihood of conversion.

Call-to-action (CTA) clarity is essential for converting interest into action. An effective ad doesn't just grab attention—it tells the viewer exactly what to do next. Whether it's "Shop Now," "Sign Up Today," or "Learn More," your CTA should be clear, direct, and easy to follow. Research by *Small Business Trends* found that CTAs phrased with action verbs boost clicks by 36%. It should also be prominently placed within the ad, so there's no ambiguity about the next step. For example, an ad for an online course might use a large, brightly colored "Enroll Now" button that stands out against the rest of the content. This makes it easy for the viewer to take immediate action, reducing the chances of hesitation or confusion.

Simplicity, Placement, and Testing

Simplicity in design and message is often the most effective approach. Ads that are too complex can overwhelm the viewer and fail to communicate the key message. A clean, simple design that focuses on one main idea is more likely to capture and hold attention. A study by *Google* suggests that simpler ads outperform complex ones, with a 61% higher click-through rate.

For instance, an ad for a new coffee blend might feature a single, striking image of the product, a brief tagline like "Awaken Your Senses," and a clear CTA. This minimalist approach ensures that the message is received without distractions, making the ad more effective.

Ad placement and timing are also critical factors in effectiveness. Where and when your ad appears can significantly impact its success. Placing your ads where your target audience is most likely to see them—whether it's on social media, search engines, or specific websites—ensures that you're reaching the right people.

Timing your ads to appear during peak engagement times, such as during a relevant event or season, can also increase their impact. For example, running ads for a summer fitness program in the spring can capitalize on people's desire to get in shape for the warmer months. Testing and optimization are the final keys to creating more effective ads.

Even the best-designed ad can benefit from A/B testing, where different versions of the ad are tested to see which performs better. This could involve testing different headlines, images, CTAs, or even slight changes in layout. The insights gained from testing allow you to refine your ads and maximize their effectiveness over time. For example, if you find that a certain image or headline consistently drives more conversions, you can apply that knowledge to future ad campaigns to improve results.

Conclusion

Creating more effective ads requires a blend of understanding your audience, clear messaging, visual appeal, storytelling, emotional connection, and strategic placement.

When these elements are combined, they create ads that not only capture attention but also convert viewers into customers. The most successful ads resonate with the audience on a deeper level, making them more likely to remember your brand and take the action you want them to take.

Call to Action

Ready to create ads that captivate and convert?

Start by focusing on your audience's needs and emotions, crafting clear and compelling messages, and using powerful visuals to tell your story. Test different approaches to see what resonates best, and optimize continuously to achieve the best results.

Get started today and watch your engagement an conversions soar!

~CHAPTER 24~

Making Visitors Feel Indebted

The Subtle Push Toward Conversion

Making visitors feel indebted is a subtle yet powerful technique in marketing that taps into the psychological principle of reciprocity. When someone receives something of value—whether it's information, a free sample, or exclusive access—they often feel a subconscious urge to give something back. In the context of online marketing, this "something" is typically a purchase, a signup, or another desired action. By strategically providing value upfront, you can gently push visitors toward conversion without overt pressure.

Offering Valuable Content for Free

One of the most effective ways to create this sense of indebtedness is by offering valuable content for free. According to a study by *Content Marketing Institute*, 70% of consumers feel closer to a company as a result of content marketing.

This could be an in-depth guide, a downloadable resource, or a useful tool that addresses a specific need or problem your audience faces. For example, if you're in the survival niche, providing a free eBook on emergency preparedness can make visitors feel grateful for the valuable information. This feeling of gratitude can translate into a willingness to explore your products further or make a purchase, as they may feel inclined to reciprocate the value they've received.

Leveraging Free Trials

Free trials are another powerful method for invoking reciprocity. Allowing visitors to experience your product or service at no cost removes the risk and gives them a chance to see its value firsthand. A report by *HubSpot* found that 57% of companies report free trials as their most effective way to convert leads into customers. When the trial period ends, many users feel a sense of obligation to continue using the service because they've already benefited from it. For instance, offering a free month of access to a premium fitness app lets users experience the benefits, making them more likely to subscribe once the trial ends. The positive experience during the trial period creates a subtle push toward making a commitment.

Creating Exclusive Offers

Exclusive offers or discounts can also make visitors feel indebted. When you present a special deal that's only available to a select group—such as first-time visitors or loyal customers—you create a sense of privilege. This feeling of being valued can prompt visitors to reciprocate by taking advantage of the offer. According to *RetailMeNot*, 80% of consumers feel encouraged to make a first-time purchase with a brand that is new to them if they receive an offer or discount.

For example, offering a 20% discount to new subscribers within their first 24 hours on your site not only incentivizes quick action but also makes them feel like they're receiving special treatment. This perceived favor can increase the likelihood of conversion, as visitors may want to reciprocate the gesture.

Personalization and Goodwill Gestures

Personalization plays a significant role in creating a sense of indebtedness.

When you tailor content, offers, or communications to the specific needs and preferences of your visitors, they feel understood and valued. This personalized approach can make them more likely to engage with your brand and take the next step. Research from *Epsilon* shows that 80% of consumers are more likely to purchase from brands that offer personalized experiences.

For example, recommending products based on a visitor's browsing history or previous purchases shows that you're paying attention to their interests. This level of attention can create a feeling of obligation to return the favor by making a purchase or taking another desired action.

Gestures of goodwill, such as offering free shipping, a small gift with purchase, or even just a heartfelt thank-you message after a transaction, can also foster a sense of indebtedness. These gestures don't have to be large to be effective; even a small token of appreciation can make a big impact.

A study published in the *Journal of Consumer Research* suggests that small, unexpected favors can significantly increase customer loyalty.

For instance, sending a follow-up email with a personalized thank you note and a small discount on a future purchase can make customers feel appreciated. This gratitude can lead to repeat business, as customers may feel compelled to return the favor.

Utilizing Referral Programs and Transparent Communication

The concept of reciprocity is also powerful in referral programs. When you offer something of value to customers who refer others to your brand—such as a discount, a gift, or exclusive access—you create a cycle of indebtedness that extends beyond the initial interaction. The original customer feels rewarded for their loyalty, and the new customer feels grateful for the referral, which can prompt them to take action.

According to *Extole*, referred customers have a 37% higher retention rate than customers acquired through other means. For example, a program that offers both the referrer and the referee a discount on their next purchase can motivate both parties to engage more deeply with your brand. Transparency and honesty in your communication can also foster a sense of indebtedness.

When you are upfront about your pricing, product limitations, or company values, you build trust with your audience. A survey by *Label Insight* found that 94% of consumers are more likely to be loyal to a brand that offers complete transparency. This trust can lead to a feeling of obligation, as visitors may appreciate your honesty and feel more inclined to support your brand.

For example, if you run a business that emphasizes ethical sourcing, clearly communicating the challenges and efforts involved can make customers feel more connected to your mission. This connection can drive them to make a purchase as a way of supporting what they believe in.

Conclusion

Making visitors feel indebted is about more than just offering freebies; it's about creating value and building relationships.

When visitors feel like they've received something of worth—whether it's information, a special offer, or exceptional service—they're more likely to reciprocate by converting. This subtle push toward conversion is effective because it's rooted in genuine human psychology, making your marketing efforts feel natural and persuasive without being forceful.

Call to Action

Now it's your turn to implement these strategies and watch your conversions soar! Start by offering something valuable to your visitors today, whether it's a free guide, a personalized recommendation, or a special discount.

Engage your audience with authentic gestures of goodwill and personalized experiences and see how quickly they feel inclined to reciprocate. Make the first move, and let your visitors respond in kind.

Your path to higher engagement and conversions begins now!

~CHAPTER 25~

Emotional AI

Leveraging Technology to Read and Respond to Customer Emotions

Artificial Intelligence (AI) has revolutionized many aspects of business, but one of its most transformative roles is in understanding and responding to customer emotions in real-time. Emotional AI, also known as affective computing, combines the power of machine learning with emotional intelligence to enhance customer interactions and improve overall user experience. By leveraging technologies such as chatbots, virtual assistants, and emotion-tracking software, businesses can create more personalized, empathetic, and effective marketing strategies.

Understanding Emotional AI: The Basics

Emotional AI refers to technologies that can detect, interpret, and respond to human emotions. This capability is achieved through various methods, including facial recognition, voice analysis, text sentiment analysis, and biometric monitoring.

By understanding a customer's emotional state, businesses can tailor their responses to better meet individual needs, thereby improving customer satisfaction and fostering loyalty.

For example, a study by Gartner predicts that by 2025, 60% of large enterprises will deploy AI in customer engagement centers to analyze emotions in customer conversations and provide adaptive responses to improve customer satisfaction and business outcomes.

This growing trend highlights the importance of integrating Emotional AI into customer engagement strategies. The ability to detect and respond to emotions allows businesses to provide more human-like interactions, even in a digital environment, enhancing the customer experience and building stronger emotional connections with the brand.

The Role of Emotional AI in Customer Service

One of the primary applications of Emotional AI is in customer service, where understanding and responding to customer emotions is crucial. Traditional customer service approaches rely on scripted responses that often lack empathy and personalization. In contrast, AI-powered chatbots and virtual assistants can analyze the tone and sentiment of a customer's voice or text to provide more nuanced and emotionally intelligent responses.

For example, if a customer sounds frustrated during a call or expresses dissatisfaction in a chat, an AI system can detect this emotion and prioritize the conversation for human intervention or offer solutions that address the customer's concerns more effectively. A survey by Forrester found that 55% of U.S. online adults are likely to abandon their online purchase if they cannot find a quick answer to their question.

By utilizing Emotional AI, businesses can reduce customer frustration and improve retention rates. Furthermore, AI systems can continuously learn from interactions, becoming more adept at recognizing patterns in customer behavior and emotional responses over time.

This ability to adapt and improve allows businesses to refine their customer service strategies and offer more personalized experiences, ultimately leading to higher customer satisfaction and loyalty.

Emotion-Tracking Software: The Next Frontier in Customer Insights

Emotion-tracking software goes beyond basic sentiment analysis to provide deeper insights into how customers feel about a brand, product, or service. By analyzing facial expressions, eye movements, and even physiological responses like heart rate and skin conductivity, this technology can gauge a customer's emotional reactions with a high degree of accuracy.

For instance, companies like Affective and Realizes specialize in emotion AI, offering tools that help businesses measure emotional responses to marketing materials, website designs, and other digital content. According to a report by Markets and Markets, the global emotion detection and recognition market size is expected to grow from $19.5 billion in 2020 to $37.1 billion by 2026, at a CAGR of 11.3%. This growth underscores the increasing demand for emotion-tracking tools that provide valuable data to enhance marketing strategies and customer experiences.

Emotion-tracking software can be particularly valuable in A/B testing campaigns. By measuring emotional responses to different versions of an ad or webpage, marketers can identify which elements resonate most with their audience and make data-driven decisions to optimize their content. This approach not only improves conversion rates but also enhances the overall customer experience by ensuring that the content aligns with the audience's emotional needs and preferences.

Personalization at Scale: The Power of Emotional AI in Marketing

Personalization is a key driver of customer engagement and loyalty. A study by Epsilon found that 80% of consumers are more likely to make a purchase when brands offer personalized experiences. Emotional AI takes personalization to the next level by enabling real-time, emotion-based customization of content and interactions.

For example, an AI-powered recommendation engine can analyze a customer's past behavior, current emotional state, and contextual data (such as time of day or location) to suggest products or services that are most likely to appeal to them at that moment. If a customer appears stressed or frustrated, the system might recommend calming products or services, such as spa packages or relaxation apps.

Conversely, if a customer seems excited or happy, the AI could suggest more energetic or celebratory products, like event tickets or adventure experiences. This level of personalization creates a more engaging and satisfying customer experience, fostering a deeper emotional connection with the brand. It also helps businesses stand out in a crowded marketplace by demonstrating that they understand and care about their customers on a personal level.

Enhancing Emotional AI with Natural Language Processing (NLP)

Natural Language Processing (NLP) is a key component of Emotional AI that enables machines to understand and interpret human language more effectively. NLP algorithms can analyze text and speech to detect not only the content of a message but also the underlying emotions and intentions. This capability is essential for creating more natural and empathetic interactions in customer service and marketing.

For instance, an NLP-powered chatbot can recognize when a customer is using sarcastic or ironic language, which might otherwise be missed by traditional text analysis methods. By understanding the nuances of human communication, the AI can respond in a way that feels more genuine and human-like. This ability to "read between the lines" enhances the customer experience and helps build trust and rapport with the brand.

Challenges and Ethical Considerations of Emotional AI

While Emotional AI offers significant benefits, it also raises several challenges and ethical considerations. Privacy is a major concern, as emotion-tracking technologies often rely on collecting sensitive data such as facial expressions, voice recordings, and biometric information. Businesses must ensure they have robust data protection policies and obtain explicit consent from customers before collecting or using such data.

Additionally, there is the risk of AI systems misinterpreting emotions or using emotional data in manipulative ways. For example, if an AI system detects that a customer is feeling sad or vulnerable, it could be unethical to exploit this emotion to push a sale. To build trust and maintain ethical standards, businesses should use Emotional AI transparently and responsibly, focusing on enhancing the customer experience rather than manipulating emotions for profit.

The Future of Emotional AI in Marketing

The future of Emotional AI in marketing looks promising, with advancements in machine learning, computer vision, and NLP continually improving the accuracy and effectiveness of these technologies.

As AI systems become better at understanding and responding to human emotions, businesses will be able to create even more personalized, empathetic, and engaging customer experiences.

According to a report by Accenture, businesses that successfully implement AI could increase profitability by an average of 38% by 2035. This potential for growth underscores the importance of integrating Emotional AI into marketing strategies to stay competitive and meet the evolving expectations of customers.

As Emotional AI technology continues to evolve, it will open up new possibilities for creating deeper connections with customers, driving engagement, and fostering loyalty. However, businesses must approach this technology with care, ensuring that it is used ethically and transparently to build trust and enhance the customer experience.

"Your emotions are the slaves to your thoughts, and you are the slave to your emotions." – Elizabeth Gilbert

Understanding and leveraging Emotional AI is about more than just technology—it's about recognizing the power of emotions in human interactions and using that knowledge to create meaningful, positive experiences for your customers.

Conclusion

Emotional AI is transforming the way businesses interact with customers, offering new opportunities to build stronger emotional connections and deliver more personalized experiences. By understanding and responding to customer emotions in real-time, businesses can create marketing strategies that resonate more deeply and drive better results.

As you explore the potential of Emotional AI for your business, consider how you can use this technology to enhance your customer interactions and build lasting relationships based on trust and empathy. Remember, the goal is not just to sell a product but to connect with your audience on a human level, making them feel valued and understood.

Call to Action

Are you ready to embrace the future of marketing with Emotional AI?

Start integrating these insights into your strategy today and watch your customer relationships—and your-business—thrive.

Conclusion

Harnessing the Psychology of Marketing for Unprecedented Success

As you reach the conclusion of this book, it's clear that understanding and applying psychological principles can revolutionize your marketing efforts. From the anchoring effect to the power of social proof, each chapter has delved into key cognitive biases and psychological triggers that influence customer behavior. By mastering these strategies, you're not just enhancing your marketing toolkit—you're fundamentally reshaping how you connect with your audience and drive conversions.

Marketing today is more than just presenting a product or service; it's about creating compelling, meaningful experiences that resonate with your audience on a deeper level. The insights provided throughout this book—whether it's using scarcity to create urgency, leveraging the power of social proof, or utilizing emotional appeals over logical ones—highlight the importance of understanding human behavior to build authentic, lasting relationships with your customers.

These strategies are not just theoretical concepts; they're actionable techniques that have been proven to work across various industries and contexts.

By applying principles like anchoring, reciprocity, emotional AI, and the power of group influence, you're equipping yourself with the knowledge to craft marketing campaigns that are not only persuasive but also ethical and customer-centric. Remember, successful marketing is not about manipulation; it's about understanding the genuine needs, desires, and pain points of your audience and providing solutions that add real value to their lives.

As Steve Jobs once said, *"The only way to do great work is to love what you do."* When you are passionate about your message, your audience can feel it, making your efforts more authentic and impactful. It's about fostering trust, building credibility, and creating a sense of community around your brand.

The landscape of marketing is ever-evolving, with new technologies and platforms continually emerging. However, the core principles of human psychology remain constant. By grounding your strategies in these principles, you're setting a strong foundation for sustainable growth, regardless of the latest trends or technological advancements. Jim Rohn wisely noted, *"Your life does not get better by chance, it gets better by change."*

Embrace the changes in the marketing landscape and use every new insight, technology, and strategy as an opportunity to adapt and grow.

Success in marketing, like in life, is a journey. *"The biggest adventure you can take is to live the life of your dreams,"* said Oprah Winfrey. Let your marketing journey be an adventure where you continually strive to innovate and create. Dream big, set bold goals, and don't be afraid to venture into uncharted territories. This adventurous spirit will not only elevate your marketing efforts but will also enrich your personal and professional growth.

Arnold H. Glasow reminds us, *"Success is not the result of spontaneous combustion. You must set yourself on fire."* Fuel your passion for marketing with the fire of continuous learning and experimentation. Let this drive be the catalyst that propels your brand and your personal growth forward.

As you implement the strategies in this book, you're not just becoming a better marketer—you're becoming a master of influence, a builder of trust, and a creator of meaningful connections.

Final Call to Action

Now is the time to put these strategies into practice. Reflect on what you've learned, identify the psychological principles that best align with your brand and audience, and start integrating them into your marketing campaigns. Test different techniques, analyze the results, and continually refine your approach.

Remember, the most effective marketers are those who are not afraid to experiment and innovate. Use the insights from this book as a launchpad to elevate your marketing efforts and achieve unprecedented success. Your audience is ready to be engaged, inspired, and moved—are you ready to lead the way?

By mastering these psychological strategies, you're setting the foundation for success, turning initial impressions into long-lasting customer relationships. The tools are in your hands; now, it's up to you to use them to their fullest potential.

Let's create marketing that doesn't just sell but transforms. Your journey to mastering the art and science of marketing starts now.

Content that Converts

NLP Techniques for Online Marketers

Neuro-Linguistic Programming (NLP) offers a significant edge in the competitive world of online marketing. At its core, marketing is about persuasion and influence, which are deeply rooted in understanding human psychology. The better you understand what drives people, the more effectively you can communicate, encourage clicks, and drive conversions.

For instance, consider online marketers in the weight loss niche who successfully convince customers to invest in a new diet plan. They don't just list features; they tap into emotional triggers and desires, like the yearning for self-confidence and better health.

This is where NLP shines. It equips marketers with tools to delve deep into customer psychology and speak the language that resonates with them. It's not just about being persuasive; it's about understanding your audience's needs and tailoring your message to meet those needs.

However, not all content is created equal. Different types of content, such as blogs, emails, sales copy, and social media posts, serve distinct purposes and play unique roles in guiding the consumer through the sales funnel. Each type of content demands a specific NLP strategy to enhance its effectiveness. Below, you'll find insights and methods to elevate your online marketing game by applying NLP across various content types.

Whether you're crafting a persuasive email or creating an informative blog, NLP techniques can help you communicate more effectively and resonate with your audience.

Understanding the Basics of NLP

Before diving into how NLP works in marketing, it's crucial to understand its fundamentals. Neuro-Linguistic Programming is a psychological approach that involves analyzing strategies used by successful individuals and applying them to achieve personal goals. It relates thoughts, language, and patterns of behavior learned through experience to specific outcomes.

NLP was developed in the 1970s by Richard Bandler and John Grinder, who believed that people are only limited by what they believe they can do. If you can change your beliefs, you can change your actions and, therefore, your outcomes.

- **Neuro** refers to the neurological processes that underpin behavior.
- **Linguistic** pertains to how language is used to make sense of the world and communicate.
- **Programming** involves setting internal codes to achieve goals.

NLP operates on several key principles. The principle of 'The Map is Not the Territory' suggests that everyone has their own perspective, and what we think is reality is just our way of navigating the world. 'You Cannot Not Communicate' implies that even non-verbal cues communicate messages, which is critical in marketing.

Lastly, 'The Law of Requisite Variety' states that the most adaptable person in a situation will control the outcome, emphasizing flexibility in the fast-paced world of online marketing.

NLP is exceptionally useful in online marketing for several reasons. First, its focus on understanding human behavior aligns perfectly with the goal of marketing. If you can understand what your customers want and how they think, you can tailor your marketing strategies to meet their needs more effectively.

Second, NLP offers tools and techniques directly applicable to content creation. For instance, understanding 'You Cannot Not Communicate' in email marketing would remind you to consider not just the content of the email but its design, layout, and timing. By grasping these fundamentals, you can gain deeper insights into your target audience and learn how best to communicate with them. This foundational knowledge paves the way for more advanced NLP applications in content-specific marketing techniques.

Applying NLP to Your Blog Content

When crafting blog posts, NLP can significantly enhance both the content and its delivery. Let's start with topic selection. The NLP principle that 'the Map is Not the Territory' is particularly useful here. Recognizing that each reader has a unique perspective allows you to tailor your blog topics to resonate with a broader audience. Techniques like customer persona modeling or sentiment analysis can help you dive deep into your audience's psyche and understand their genuine interests.

For example, instead of writing a generic post about "The Importance of SEO," you might address the emotional and psychological barriers that prevent people from investing in SEO services. A blog post like "Overcoming the Fear of SEO Complexity: A Guide for Beginners" uses NLP principles to address emotional resistance. The topic is not just about SEO; it's about the psychological hurdles related to it, making it more relatable to readers who feel overwhelmed by SEO.

In terms of writing techniques, NLP offers powerful tools like language patterns and storytelling. One effective NLP language pattern is the "embedded command," a way of embedding a suggestion within a sentence to subtly guide the reader towards a particular action. For instance, "As you read through this guide, you'll naturally start to see the value of SEO in a new light." Here, the embedded command is to "start to see the value of SEO," subtly influencing the reader's perception. Storytelling is another potent NLP technique.

Instead of merely presenting facts or tips, you tell a story that emotionally engages the reader. In the weight loss niche, rather than a straightforward guide on *"Losing 10 Pounds,"* frame the blog post as a journey—highlighting an individual's initial struggles, the moment they found your weight loss program, and their subsequent transformation. This narrative approach immerses readers in an experience, not just information, making your content more compelling.

By applying NLP techniques to your blog content, you can create posts that go beyond conveying information—they connect, persuade, and ultimately convert.

Leveraging NLP for Your Affiliate Product Reviews

Affiliate reviews offer a unique opportunity to persuade readers to invest in third-party products or services based on your recommendation. Trust is your most critical asset in this scenario. Readers won't take your advice unless they find you credible. An NLP strategy to build trust is "mirroring," which involves reflecting the reader's style, tone, or language to create rapport.

For example, if you're reviewing a weight loss supplement and your audience is frustrated with yo-yo dieting, use language that mirrors these emotions: *"I get it; you're tired of diets that promise the world but deliver nothing."* This demonstrates empathy and understanding, making your review more relatable.

Offering a balanced perspective by presenting both positives and negatives can also build trust. This honest approach can make readers more likely to trust your final recommendation. For instance, you could say, *"While this supplement has helped curb my appetite, it took about two weeks to see noticeable results."*

This statement sets realistic expectations, mirroring the reader's likely desire for quick results. NLP also offers several persuasion techniques to encourage the reader to make a purchase. One effective technique is "future pacing," which involves getting the reader to imagine a future where they enjoy the benefits of the product.

For example, in a review for a digital marketing tool, you might write, "Imagine a month from now, your website traffic has doubled, and your conversion rate is the best it's ever been. This tool can help you get there." This paints a vivid picture of the positive outcomes the reader could experience, making them more inclined to consider the product.

Another technique involves "anchoring," which attaches a positive emotion to the action you want the reader to take. For instance, saying, "The moment I clicked 'buy,' I felt a sense of relief knowing I was taking a positive step toward a healthier me," can encourage the reader to seek the same emotional experience. Applying these NLP techniques can transform your affiliate reviews from mere rundowns of features into persuasive content that builds trust and guides the reader toward making a purchase.

NLP to Boost Email Marketing Conversions

Emails are a direct line of communication with your audience, making them a prime candidate for NLP techniques to boost engagement and conversions. The first impression a recipient gets is often from the subject line, which is crucial for drawing people in.

In NLP, 'curiosity gaps' are particularly effective for subject lines. This involves creating enough intrigue to make the reader curious but not so much that their curiosity is satisfied. For instance, if promoting a new marketing course, a subject line like *"Unlock the Secret to Triple Your Traffic"* teases a benefit or result, piquing curiosity and tapping into a specific desire.

In the body text, several NLP techniques can be used to engage and persuade. "Pattern interrupts," for instance, are used to break the reader's unconscious flow and grab their attention.

Starting your email with a question that challenges the reader's beliefs can engage them on a deeper level. A question like, "Have you ever wondered why some blogs explode in popularity while others don't?" prompts reflection and makes the reader more receptive to the content that follows. Once you have their attention, "embedded commands" can subtly nudge the reader toward the desired action.

For example, in an email promoting a weight loss product, you might write, "As you discover the benefits of this supplement, you'll find yourself clicking the 'Learn More' button below." The command "click the 'Learn More' button" is embedded within a larger, less direct sentence, gently guiding the reader.

"Social proof" is another compelling NLP technique in email marketing. Mentioning how many people have benefited from your product can serve as an endorsement and trigger a psychological desire to join the 'winning team.' A line like, "Join the 5,000+ marketers who have revolutionized their campaigns" can motivate readers to take action.

By incorporating these NLP techniques, your emails can become powerful tools for engaging your audience more meaningfully and increasing conversions.

Mastering Copy with NLP Persuasive Skills

Sales copy is often the final touchpoint before a potential customer decides to buy, making it crucial to craft persuasive content. NLP can enhance sales copy by informing what you say and how you say it.

The headline is the first point of contact and often determines whether the reader will continue reading. The NLP principle of 'immediacy' suggests that people are more likely to act when they perceive immediate benefits. Instead of a headline that says, "The Ultimate Guide to Weight Loss," use one that offers a clear, immediate benefit, such as "Lose Your First 5 Pounds in Just One Week!" This headline grabs attention and creates urgency, encouraging the reader to engage further with the content.

For the storyline, organizing information using a narrative arc can keep the reader engaged from start to finish. NLP emphasizes the power of storytelling to engage emotions and capture the reader's imagination. If you're selling a digital marketing course, your sales copy could start with a relatable problem ("Are you tired of wasting money on ineffective marketing strategies?").

It would then introduce a turning point ("That was me until I discovered these simple yet powerful techniques") and conclude with a resolution ("Now, I want to share these proven methods with you"). This creates a relatable journey that potential customers can envision themselves being a part of, thereby increasing their emotional investment in the product. When listing product benefits, it's crucial to use NLP techniques that appeal to different sensory modalities (visual, auditory, kinesthetic) to make the content more engaging.

For example, in a list of features for an online marketing tool, you could include benefits that cater to different senses: "See your traffic surge with real-time analytics" (visual), "Hear the cha-ching of increasing sales" (auditory), and "Feel the satisfaction of finally having a successful campaign" (kinesthetic). This multi-sensory approach makes the content more relatable and memorable.

The Call to Action (CTA) is where the power of NLP really comes into play. One effective technique is the use of "command verbs" to prompt a specific action. Instead of a generic CTA like "Get Started," opt for something more direct and action-oriented like "Start Your Free Trial Now" or "Download Your Free Guide Today." These CTAs not only tell the reader exactly what to do but also what they will gain by taking the action, making it more likely they will follow through.

By employing these NLP techniques, your sales copy can go beyond merely informing to actively engaging and persuading the reader, thereby increasing the chances of conversion.

Attract Attention on Social Media with NLP Strategies

Social media platforms provide a unique opportunity for marketers to directly engage with their audience, making them an ideal space to apply NLP techniques tailored to different platforms and objectives.

Different platforms require different NLP approaches. For example, YouTube is conducive to longer content, allowing for rapport-building through techniques like "mirroring" and "pacing." Conversely, platforms like TikTok, with their shorter content formats, benefit more from immediate-impact NLP strategies like "pattern interrupts" or "immediacy" to capture attention quickly.

On platforms like Instagram and Facebook, where visual and text content coexist, NLP techniques such as visual anchoring and storytelling can be highly effective. For example, pairing a captivating image that evokes feelings of freedom or success with a short narrative or an inspirational quote can induce those feelings in your audience.

Audience engagement on social media often revolves around generating likes, shares, and comments, which can be influenced by NLP. "Triggers" can be employed to elicit specific responses, such as ending a post with a question that encourages comments.

For instance, "What's your go-to strategy for increasing site traffic?" invites interaction and opens the door for deeper conversations about your products or services. For shares and likes, using affirmations or motivational quotes that resonate with your audience's values can encourage them to share your content. A post like, *"Success is not the key to happiness; happiness is the key to success. If you love what you are doing, you'll be successful,"* can inspire shares from those who agree with the sentiment.

Influencing behavior is the ultimate goal of social media marketing. Guiding your audience to take specific actions can be achieved through NLP techniques like "directional cues," both verbal and visual. Phrases like "Swipe up to learn more" or "Click the link in our bio," paired with an arrow or visual cue, can effectively guide user behavior.

Another way to influence action is through "social proof," particularly potent on social platforms where user interaction is visible. Posts showcasing customer testimonials or user-generated content can positively influence how your audience perceives your brand, encouraging them to make a purchase or try your service.

By adapting NLP techniques to the unique characteristics and user behaviors of each social media platform, you can engage your audience more effectively and guide them toward desired actions.

Taking NLP to the Next Level

For those who have mastered the basics of NLP and want to deepen their impact, exploring advanced NLP techniques can unlock new levels of engagement and conversion.

Meta Programs are the filters through which people process information and make decisions. These ingrained mental programs guide behavior and reactions and can be complex but understanding them allows you to segment your audience more effectively and tailor your content to resonate with different types of decision-makers.

For example, some individuals are more motivated by moving toward goals, while others are driven by moving away from problems. Knowing this, you could create two versions of the same marketing message: one focusing on the gains *("Achieve financial freedom")* and another on avoiding pain *("Say goodbye to financial stress")*.

NLP Strategies are step-by-step mental processes that people use to achieve particular outcomes. For marketers, understanding these strategies can provide a blueprint for how a person makes a purchase decision. Aligning your sales funnel with these steps increases the likelihood of conversion.

If you discover that your customers often research online reviews before making a purchase, you could focus on accumulating positive reviews or creating content that showcases customer testimonials.

NLP Modeling involves studying successful individuals to understand their beliefs, strategies, and mental models that lead to their success. These can then be replicated for similar results.

In marketing, this might involve analyzing highly successful campaigns to understand which NLP techniques were used.

For example, if a successful weight loss product campaign heavily utilizes storytelling, you might deduce that emotional engagement is a key factor in its success and incorporate similar storytelling elements into your campaign. NLP modeling goes beyond mimicry; it involves adapting foundational NLP principles to your unique context and offerings. By understanding advanced NLP techniques, such as Meta Programs and NLP Modeling, you can craft more targeted and effective marketing strategies, better engage your audience, and guide them more efficiently through their buying journey.

Conclusion and Call to Action

NLP offers a powerful set of tools for online marketers looking to connect more deeply with their audience, improve engagement, and drive conversions. By understanding and applying these principles, you can transform your marketing strategies from good to great, creating content that not only informs but also persuades and converts.

"Marketing is no longer about the stuff that you make, but about the stories you tell." – Seth Godin.

Using NLP, you can tell stories that resonate, create content that engages, and build relationships that last. As you continue to explore the world of NLP, remember that the most successful marketers are those who are willing to adapt, innovate, and continuously learn. Take the next step in mastering NLP techniques for your marketing efforts. Start by reviewing your current strategies and identifying where NLP principles can enhance your approach. Experiment with different techniques, analyze the results, and refine your methods.

Your journey to becoming a more persuasive, effective marketer starts now.

Leverage the power of NLP and see how it can transform your marketing strategy and boost your conversions.

Are you ready to take your marketing to the next level?

Start today!

Suggested Book

Visit:

UnspokenInfluenceBook.com

Made in the USA
Columbia, SC
16 September 2026

0788fb28-6860-44eb-ae7d-15b6b865789eR01